DIALOGUE
WITH

C.G.
JUNG

DIALOGUES IN CONTEMPORARY PSYCHOLOGY SERIES

Richard I. Evans, Series Editor

DIALOGUE WITH

C.G. JUNG

WITH

Richard I. Evans

PRAEGER

PRAEGER SPECIAL STUDIES • PRAEGER SCIENTIFIC

Library of Congress Cataloging in Publication Data

Evans, Richard Isadore, 1922-
    Dialogue with C. G. Jung.

    (Dialogues in contemporary psychology series)
    Reprint. Originally published: Jung on elementary
psychology. 1st ed. New York : Dutton, 1976.
With a new introd.
    Bibliography: p.
    Includes index.
    1. Jung, C. G. (Carl Gustav), 2. Psychoanalysis.
I. Jung, C. G. (Carl Gustav), 1875-1961. II. Title.
III. Series: Evans, Richard Isadore, 1922-   .
Dialogues with notable contributors to personality
theory; [v. 1]
BF173.J85E9 1981   150.19′54   81-15371
ISBN 0-03-059927-X   AACR2

BF
173
.J 85
E9
1981

Published in 1981 by Praeger Publishers
CBS Educational and Professional Publishing
a Division of CBS Inc.
521 Fifth Avenue, New York, New York 10175 U.S.A.

©   1964 & 1976 by Richard I. Evans and
    1981 by Praeger Publishers

123456789 145 987654321
Printed in the United States of America

To
my lovely wife
and children

Introduction to the Praeger Edition

The first edition of this book was published in 1964. Since then, C. G. Jung has been increasingly viewed as a major contributor to the behavioral sciences, who truly anticipated the importances of life span development. As current theory and research in psychology is increasingly concerned with crises of middle age and aging, one can examine Jung's perspective and see how he considered the importance of these stages of development in his concept of individuation. The recent interest in sociobiology would have been no surprise to Jung as an extension of his notion of archetypes. In the present volume the reader has an opportunity to examine such Jungian conceptions against the background of current ideas. His often surprising and candid views of his own work, and of various figures in the history of psychoanalysis, provide a rare glimpse of the one contributor to psychology and psychiatry who truly transcended disciplinary lines. With this new edition, readers who in the past might have been dissuaded from reading Jung will have an opportunity to be introduced to his ideas through the spontaneity of our conversation, which leads him through a thoughtful examination of his major contributions.

CONTENTS

II. THE UNCONSCIOUS   63

III. MAJOR THEORIES AND CONCEPTS   89

IV. ISSUES AND INSIGHTS   117

V. CONCLUSION   161

Acknowledgments

In the long process involved in filming the dialogues with Carl Jung in Europe and transcribing them for the present volume, I am indebted to a great many individuals. Though space prohibits mentioning everyone who so kindly assisted, I wish to express my appreciation to at least some of those people who contributed. Dr. John W. Meaney, now of Notre Dame University, who functioned superbly in the demanding roles of producer-director-cinematographer for the original films, and without whose support the entire project would have been impossible, must be cited most prominently.

The special consideration and assistance of Frau Aniela Jaffé, Dr. Jung's secretary-assistant and now an author in her own right, proved invaluable in helping us arrange and successfully complete the interviews with Dr. Jung. The encouragement and support of Dr. Joe Wheelwright, the eminent Jungian psychiatrist at the Langley-Porter Neuropsychiatric Institute in San Francisco, provided the critical personal endorsement we needed to elicit the cooperation of Dr. Jung.

The willingness of the Swiss Federal Institute of Technology in Zurich to provide physical facilities for the interviews is also greatly appreciated.

Thanks are accorded psychology graduate students Albert Ramirez and Gary Blank for their assistance. Miss Joy Byrne's skill and imagination as an editorial assistant helped me throughout in the preparation of the manuscript for the first edition, and for her efforts I am most grateful, as I am to psychology graduate student Bettye Earle Raines who served most effectively in this capacity during the preparation of the second edition.

In the process of preparing this material for inclusion in a forthcoming volume, a group of editors of the *Collected Works of C. G. Jung,* including William McGuire and the late R. F. C. Hull, compared the original transcriptions of the discussions to the soundtracks of the films, making textual corrections, particularly in the areas where Jung's speech and use of certain foreign phrases made transcription difficult. Their kindness in making these corrections available to us is much appreciated.

We are grateful for the grant from the Fund for

the Advancement of Education. The Fund's tolerance in allowing us to deviate from an original plan, which would have led simply to the recording of lectures, provided the financial assistance and latitude without which this project could not have originated.

Introduction
to the
Second Edition

In 1957 I was granted a series of interviews with Carl Jung, which were published, under the title *Conversations with Carl Jung,* in 1964. Since that time several developments have occurred which make the publication of this second edition necessary and desirable.

The most important of these has been the immense growth of interest in Jungian psychology. In the early 1950s I was one of very few social-personality psychologists who began teaching Jungian theory as part of a personality psychology course; at that time Jung was more likely to be dismissed as a mystic. It

1

is no wonder that Jung's first question to me when I met him in the garden of his home in Küsnacht that summer of 1957 was, "Why do you American psychologists hate me so much?" It would have amazed him to know that the very first interview (Evans, 1967)[1] in that new American magazine, *Psychology Today,* was taken from the discussions we completed that summer; that his ideas would be a subject of interest as presented by American psychologists on popular nationally televised "talk" shows, as they have been on several occasions; or, in the more academic realm, that his theories would be presented in virtually all college psychology courses, if not always in depth, at least with a sense of their historical significance.

The historical importance of these discussions has also become increasingly apparent. Among Jung's few filmed appearances, these are perhaps the only ones in which he was asked specifically to address himself to elementary concepts and to beginning students. Though his outward manner appeared receptive to this approach, it would not have surprised me to learn that he really felt that this American psychologist was "putting him on a leash" with these simple questions that restricted a more profound discussion. In fact, *Time* magazine accompanied the story that described our discussions with a unique photograph of Jung wearing a microphone around his neck, and captioned, "Jung on a leash."

As I prepare the second edition, however, the

---

[1] Notations in parentheses refer to listings in References.

original purpose of these discussions still seems to me valid; that is, to introduce the beginning student to Jung's ideas; to stimulate those readers who had been discouraged from reading the original works because of their alleged obscurity, complexity or mysticism; and to provide for the more advanced Jungian scholar a more intimate glance at Jung's thought processes as he reacted *spontaneously* to an orderly sequence of questions.

Following the recorded interview on each of the four days which Jung allowed me, he clarified some of the points he had made, and this is reflected in the edited version, particularly when he had not clearly understood a question. Because of his difficulty in hearing, I stated my questions in a bit more detail than would ordinarily have been necessary, and some of this reiteration was deleted to enhance written communication. As an appendix to this edition, a transcription of the actual soundtrack of the filmed discussions is included. This transcription has been carefully compared to the recorded soundtrack by a group of Jungian scholars, including William McGuire, the late R. F. C. Hull, Aniela Jaffé, Marie-Louise von Franz and Barbara Hannah, and their suggestions, particularly in those areas where Jung's accent and use of either foreign or highly technical terms made transcription difficult, have been integrated into the discussions.

Some minor differences between the actual transcriptions and the final edited version reflect an effort to bring together the major concepts in an orderly fashion for the introductory student. Others reflect the highly informative, but unfortunately unrecorded comments evolving out of more informal discussions following the recording sessions. In spite of the difficulties encountered in this venture, it was reassuring that readers of the first edition who were acquainted with Jung, such as Henry Murray and Aniela Jaffé, seemed to feel that the essence of Jung and his ideas had been successfully captured.

The new title for this edition reflects the fact that my discussion with Jung covered basic psychological issues, not metaphysical concepts. It would have been literally impossible to do justice to most of Jung's spiritual and metaphysical ideas in the limited time he allowed for our discussions. In fact, I doubt that in all of his published works are the subtleties of his own "active imagination" fully realized.

The fine response to this discussion with Jung encouraged me to continue a series of filmed and audiotaped discussions with the world's notable contributors to psychology, including Erich Fromm, B. F. Skinner, Erik Erikson, playwright Arthur Miller, Gordon Allport,[2] Jean Piaget, Carl Rogers, Konrad Lorenz, R. D. Laing, and Nikolaas Tinbergen. Films[3] produced from these discussions are now being used in conjunction with various courses in over 300 American universities, and a series of books has been published based on transcriptions of the dialogues.

[2] *Gordon Allport: The Man and His Ideas* received the 1971 American Psychological Foundation Media Award in the book category.
[3] The films are distributed by Macmillan Films, Inc., 34 MacQuesten Parkway South, Mt. Vernon, New York 10550.

Designed as an innovative teaching device, the series was supported by a grant from the Fund for the Advancement of Education and is being continued under a current grant from the National Science Foundation. A basic purpose of the project is to produce for teaching purposes a series of films that introduce the viewer to outstanding contributors to the field of psychology and human behavior. We hope that these films may also serve as documents of increasing value to the history of the behavioral sciences.

The books in this series are based on edited transcripts of the filmed dialogue, including, in some instances, audiotaped discussions as well as the contents of the films. These dialogues are designed to introduce the reader to the contributor's major ideas and points of view, conveying through the extemporaneousness of the dialogue style a feeling for the personality of the contributor.

Since the questions in this series reflect many of the published writings of the interviewee, it might be expected that a comprehensive summary of his work is evoked. However, the selectivity necessary in developing the questions within a limited time interval does not always provide the basis for an inclusive summary. In fact, we are hoping to present a teaching technique leading away from the trend observed among many of our students today to become increasingly content with only secondary sources for gaining information concerning major contributors to various disciplines. The material—films and books— resulting from our dialogues provides a novel "original" source exposure to the ideas of outstanding persons, which in turn may stimulate the viewer or reader to go back to the original writings which develop more fully the ideas presented through our discussions.

It is my intention that these dialogues offer a constructive, novel method of teaching, with my role as interviewer being neither the center of focus nor critical challenger. The purpose of this book will be realized if I am perceived as providing a medium through which our distinguished interviewees can express their views. It is within the spirit of these teaching aims that our contributors so generously participate, as is evident, for example, in a letter from the late Carl Jung, reproduced in this volume. Using such sessions primarily as a background for critical examination of the views of the participants must be left to another type of project, since even if this critical set were to be emphasized in my questioning, it might be difficult to introduce the reader to the contributor's views and criticize them as well, within the limited time commitment. I expect that some of the participants who agreed to work with us on this project would not have done so if they had sensed primarily a critical attack on their work.

As was the case with subjects of other books in the series, it is hoped that the dialogue presentation allows the reader to be introduced to, or reexamine some of Jung's ideas through a relatively extemporaneous situation. It must be pointed out, however, that in his own writings, Jung expressed himself in his own unique style, and he had the opportunity to rewrite and polish until he deemed the finished product satisfactory. In the spontaneity of our discussion, he was called upon to develop his ideas extemporaneously. I hope this element of spontaneity may present more of the "man behind the book" while losing none of the ideas central to his thought.

Because preservation of this naturalness of communication is essential to the purposes of each volume in this series, few liberties have been taken with the basic content of Jung's responses to my questions. The dialogue presented here duplicates insofar as possible the tenor of the exchange between Dr. Jung and myself as it actually took place. In spite of the editing that was necessary, it was a pleasant surprise to review our hours of discussion and realize how few deletions and alterations were required. I hope that this dialogue makes available to the reader some reactions not readily obtainable from Jung's more traditional didactic presentations or from the secondary sources on his work in the literature.

Rather than attempt to summarize all of the major concepts presented here, I shall take the liberty of briefly presenting frameworks which I find valuable in teaching personality theory to students.

There are three frameworks around which I believe current approaches to personality psychology can be analyzed in order to help locate any theoretical position within the matrix of general personality theory. These frameworks are really descriptive approaches to the understanding of personality, which develop theoretically from basic orientations focusing around the unconscious-oriented biological-determinism, the social/environment-oriented cultural-determinism, or experience-oriented self-determinism.

One group of contributors, emphasizing unconscious-oriented biological-determinism, has been considered more or less traditionally psychoanalytical, and includes such writers as Hans Sachs and Ernest Jones, as well as Freud himself. This group has been characterized as emphasizing what Freud called repetition compulsion, a concept that maintains that the first five years of life, which are strongly influenced by biological propensities, are very important in human development because they set the stage for and determine a lifestyle that is manifested continuously throughout the individual's lifetime. Central to this postulate is the notion of the Oedipal complex. Another important aspect of traditional Freudian theory was brought out by Ernest Jones when he said, in a discussion with me, "Well, man is, after all, an animal." Some people think that this is a cynical view, although Jones denied that Freud was inordinately cynical. Interestingly, the increasing impact of ethology on contemporary psychology has reopened this issue in a new vein, somewhat more palatable to psychologists. Freud's earliest picture of man is that of an organism dominated to a large degree by its id, the animal, biological side of him—against which the ego, the conscious, the self of man, is fighting a tough battle. This view, articulated in many of Freud's early works, was also accepted by many of his early followers. With Freud, they believed that the center of man's motivation is the sexual libido, which to them was a manifestation of the dominant animal aspect of man. Although Freud, in his later work, began to emphasize other aspects of man's makeup, many thinkers continue to perceive the classical psycho-analytical position in terms of Freud's early views. This description is probably a vast oversimplification of Freud's view, as Fromm and Erikson, for example, implied in my discussions with them.

Another group, the neo-Freudians, has placed more stress on the effects of social-environmental cultural influences on man's development. To the neo-Freudians, the early Freudians emphasized too much the notion that the instinctual animal nature, the repetition compulsion, and a general biological pat-terning of early developments are found *universally*, and that these elements dominate man's nature. The neo-Freudians take exception to this concept of uni-versality. They believe that man is primarily a product of the specific kind of culture in which he lives, and that learning plays a much more important part than does biological patterning in the development of personality.

The late Karen Horney (1937), for example, became so disturbed by many notions of the biological orientation of the early Freudian position that she broke away from the orthodox position and developed the view that people are shaped to a significant extent by the society with which they must cope when they deal with the anxieties of reality. She considered this anxiety produced by societal pressures more impor-tant in shaping us than our anxiety about overcoming our basic biological animal natures.

Although Fromm does not like the label neo-

Freudian he, too, certainly takes exception to the emphasis on the Oedipal situation so central to Freud's "biological unfolding" view of man's development.

Virtually all contemporary American psychologists have attempted to place man within his social milieu, in the belief that it constitutes the essential force in shaping personality. The neo-Freudians, however, still did not pay adequate attention to the principles of learning that are necessary to account for the shaping influence of the social environment.

It must be kept in mind that related to any theoretical discussion of determinism and personality theory, the behavioristic orientation may still be perhaps the most significant theoretical reference group for American academic and research psychologists. As the leading contemporary exponent of this view, B. F. Skinner interprets not merely cultural influences in a broad sense, but the immediate environment in a narrow sense as being the significant shaping force on the individual. As environment is controlled experimentally, and even in clinical situations, to modify behavior in a desired direction, very few assumptions are made concerning the internal workings of personality. Skinner supplied a cogent model of the process of shaping and controlling behavior, strongly influencing contemporary thinking.

In direct contrast to Skinner, Carl Rogers also formalized a set of learning principles which he believed to be useful, but not controlling. A great deal of thought today continues to reflect the greater concern for man's individuality and self-responsibility than is found in either biological or cultural determinism. This concern is seen in the work of Rogers, Maslow (1954) and May (1950); in that of the more theoretically-oriented psychologists such as Allport and McCurdy (1961); and such philosophers as Husserl (1952) and Heidegger (1959), all of whom are concerned with the autonomy of the self.

Somewhere between the neo-Freudians and the traditional Freudians there is a group of three significant individuals whom we might describe as Freudian dissentients; for although each worked closely with Freud, each subsequently broke with him, or was repudiated by him, for one reason or another: Carl Jung, Otto Rank, and Alfred Adler.

Adler's (1927) early work placed the primary emphasis on the social man, and it might be said that Adler set the stage for the emergence of the neo-Freudian group. In a different direction, although many of his ideas about early biological concepts were in agreement with Freud's, Rank's preoccupation with the "will" and its development of autonomy introduced a type of self-determinism that Freud did not emphasize. Rank had a marked influence on Rogers, for example.

As becomes apparent in this dialogue, Jung moved away from Freud's basic tenets, while retaining Freud's idea of the unconscious, expanding it into a race or collective unconscious and the individual unconscious. He incorporated into the collective conscious Freud's early notion of archetypes, developing this concept beyond Freud's postulation. However, with his central conception of individuation, Jung also moved away from the emphasis on biological determinism. Jung, perhaps more profoundly than either Adler or Rank, turned toward the idea of the development of an ultimately self-determined spiritual being that transcends the biological forces acting on man. This led him to consider many metaphysical concepts which have been difficult for present-day scientific psychology to accept.

This shift from biological, to social, to self-determinism as a means to account for the human predicament is really, however, a matter of emphasis. In spite of the claims of the extreme champions of each of these positions, one must recognize that probably no simple conception of determinism, whether it be biological, social or radical free-will-self-responsibility, can adequately and completely account for the complex behavior of the individual or the group. Interestingly enough, Jung reflects all of these trends at various points in these interviews, showing some tolerance toward a composite of these influences.

For example, visible in many of Jung's observations is that portion of his focus which, by his own admission, is quite compatible with the biological determinism of Freud. At another point, however, when we were attempting to determine whether Jung necessarily would emphasize the importance of historical analysis in understanding the individual, he indicated that he could see the virtue of a historical or field theoretical analyses as well. In fact, Jung's responses suggested a surprisingly well-balanced acceptance of the importance of both types of analyses. This is particularly interesting in view of the priority that Freud had bestowed upon primarily historical determinants of personality, as well as the obvious historical deterministic implications of Jung's collective or race unconscious. At several points in the dialogue, Jung's reference to and understanding of cultural determinants was apparent through his descriptions of the various cultures he had studied, though many of his conclusions are hardly congruent with the ideas of present-day cultural anthropologists.

The apparent compatibility to Jung of a theory of universal archetypes joined in each individual's collective unconscious, coupled with the molding of the behavior precipitated by them showing differentiation, at least in part due to variations in cultural patterning, may be one of the most interesting insights he provided in these interviews.

Finally, his concern with the nature of an intrinsic self and the individuation process is certainly consistent with the current interest in self-deterministic perspective in personality theory. This may explain the renewed interest in Jung and his work that is becoming evident in many quarters.

Prologue to a Challenging Venture

I shall attempt here to trace in some detail the challenging series of events that preceded the interviews with Dr. Jung in August 1957. This section also includes some of my impressions and perceptions at that time, of the actual process of completing the interviews, including certain personality insights gained by even limited contact with so notable a figure as Carl Jung.

The idea of filming a series of interviews with Dr. Carl Jung, when it first occurred, seemed to be only a remote possibility. We knew that Dr. Jung had

14

been approached unsuccessfully by several commercial television and film producers in the past. Yet it seemed worth a try.

The University of Houston had received an $18,-700 grant from the Fund for the Advancement of Education to explore some new dimensions in university instruction. Dr. John Meaney, who at that time was director of the University of Houston Radio and Television Film Center, was involved with this project when he approached me concerning the possibility of utilizing these funds in some project involving Psychology. We began to discuss possible approaches that could be pursued if a pilot project in Psychology was initiated with these funds.

For years, like many other university professors, I had become increasingly aware of the tendency among great numbers of students to become less interested in reading the original writings of great contributors. They were becoming content merely to read "predigested" secondary sources which often did injustice to the intentions of these significant thinkers. For example, it always amazed me to find that a surprisingly large number of even advanced psychology students had never actually read Freud's original writings, but had instead read *about* Freud through the words of other writers. It seemed that there was a need to motivate the students to look directly at the original writings of such innovators as Freud, if they were to become truly informed and appreciative of their ideas. It then followed that a challenge lay in developing, with our grant funds, a stimulating technique that would encourage the students to pursue such primary contact with the ideas of important contributors.

Simply making films of lectures by these individuals, wherein they orally presented the same material about which they had already written, did not appear to be the most effective means for our purposes. The potential pitfall in this method is readily discernible in many college television courses. Specifically, the intellectual, in lecturing as in writing, tends to become somewhat pedantic, thereby losing the interest of his audience. It occurred to me that perhaps one technique that might be utilized to avoid this undesirable, imminent possibility was the filmed interview. Through use of the interview, the contributor could present his ideas in an atmosphere of spontaneity which would tend to "humanize" him, providing for the student a more pleasant and stimulating experience than is often allowed by the neutrality of the formal lecture.

The interview, of course, has long been used as a technique in such fields as journalism, law, psychotherapy, counseling and casework, and obviously is a fundamental device in our culture for gaining insight into other people and their ideas. Why could not carefully planned and filmed interviews be set up with eminent psychologists for instructional purposes? The student, through the interviewer, could be systematically introduced to a great contributor's point of view. We liked the idea.

A course that I had offered for many years, "Approaches to Personality," appeared to be a likely vehicle for such an effort, so all that remained was to find a manner in which to launch this technique in as dramatic a fashion as possible. Thus, the idea of interviewing Dr. Carl Jung, the only surviving member of the "big three"—Jung, Freud, and Adler—originated.

Most individuals were very skeptical of our chances for success when we announced that we were going to contact Jung and pursue the possibility of going to Zurich to film a series of interviews to launch our teaching-interview project. The idea of interviewing Jung was too appealing, however, so we prepared to write to him.

The framing of this letter became an unusual task. One asks onself, if he has never met the man, how Jung's interest in such a project might best be solicited. Also, when one has spent many years of one's life studying personality theory and has come fully to appreciate the historical importance of Jung in the psychoanalytic movement, the task of writing to him takes on a certain air of excitement.

In order to gain perspective for this task, I decided to contact Dr. Joe Wheelwright, a prominent Jungian psychiatrist at the Langley Porter Neuropsychiatric Institute in San Francisco. He had had considerable contact with Dr. Jung, and could be of much assistance to us as a liaison in our efforts to secure the cooperation of Dr. Jung. Thus, it was with gratification that my colleague and I received not only Dr. Wheelwright's professed interest in the proposed project, but permission from him to mention his support of it in the letter subsequently dispatched to Professor Jung, requesting his participation in the initial interview of our series. That letter read as follows:

April 2, 1957

Professor Doctor C. G. Jung
Zeestrasse, 231
Küsnacht bei Zurich,
Switzerland

Dear Professor Doctor Jung:

A prominent United States foundation, The Fund for the Advancement of Education, has awarded us a small grant which will make it possible to begin filming for the use of undergraduate students a psychology course series of lectures and discussions.

In planning a course in psychology on film, it occurred to us that the presence on film of some of the truly great men in psychology would be an inspiration to our American psychology students. Naturally, the first name that came to our minds was yours. We have long been interested in your work, and your presence on film would, in our opinion, add appreciably to the learning of our students.

If you would be willing to participate, we could fly to Switzerland to do the filming at your convenience. We would not request a great deal of preparation such as formal lectures would require, but rather, we would ask you to participate in a series of four informal interviews. We would, of course, submit the topics to you in advance, and in fact, would welcome your advice in choosing them. This would allow you to reflect fully the many interesting facets of your work. To avoid imposing on your time, these interviews could be spaced so that they could be filmed over a week's time or longer.

We would plan to spend a week or more in Switzerland, and, if it will fit into your schedule, could arrive on or around August fifth.

Dr. Joe Wheelwright, with whom we spoke concerning this matter, wishes to express his encouragement to you to work with us on these films. He shares our belief that it would be of great educational value to our psychology students, not only in this university but throughout the United States. Copies of the films could be made available to colleges everywhere in the United States.

Dr. John Meaney, Director of the Radio-TV Film Center at this university, as recipient of the grant, would produce the four films. He has produced many stimulating educational series for professional groups and for educational television. From my own experience working with him, I find him a most sympathetic and understanding student of psychology; so his work, I'm certain, would achieve the best possible results.

If you will permit us to do so and will suggest an appropriate amount, we shall be pleased to arrange for payment of an honorarium to you for your participation in these four films.

We are looking forward hopefully to a reply from you concerning this matter.

Cordially yours,

Within ten days we received the following reply from Dr. Jung:

Prof. Dr. C. G. Jung   KÜSNACHT-ZURICH
                               Seestrasse 228

April 1957

Prof. Richard I. Evans, Ph.D.
University of Houston
Cullen Boulevard
Houston 4, Texas

Dear Prof. Evans,

I am inclined to meet your request, if you can limit yourself to four interviews on consecutive days, beginning on August 5th about 4 P.M.

As to the nature of your questions, I prefer your initiative. I would not know in what aspect of psychology you are particularly interested, I also cannot assume that our interests are the same. An interview should not last longer than one hour at the most, as I easily get tired on account of my old age.

Since I am not informed about the size of your grant, I should like you to tell me frankly what you intend in the way of an honorarium.

I hope you are sufficiently aware of the unreliability imposed upon me by my age. Whatever I promise is necessarily subjected to the ulterior decision of fate that can interfere unexpectedly.

                                Sincerely yours,

I'm sure that you can imagine the delight with which we greeted this reply and the haste with which we proceeded in the direction of further planning. The following correspondence between Dr. Jung, myself, and Dr. Jung's secretary, Aniela Jaffé, is self-explanatory and traces the lines of events that led us to set up a firm appointment for four days in August of 1957.

April 18, 1957

Professor Dr. C. G. Jung
Seestrasse 228,
Küsnacht-Zurich, Switzerland

Dear Professor Doctor Jung,

We were all delighted to receive your letter of April 12. On the day that your letter arrived, it happened that we were discussing some of your contributions to personality theory in my Psychology of Personality class, and when I read your letter to the class, it was, indeed, a dramatic note.

The dates that you indicate that you can see us for the purpose of filming interviews, August 5, 6, 7, and 8, are just fine. Dr. Meaney and I would probably arrive in Zurich a few days prior to this, of course.

In looking over our budget, an honorarium in the amount of five hundred dollars would appear to be feasible. Does this seem to be sufficient? If not, please let us know and we shall make every effort to make some adjustment.

With respect to the content of the interviews and the kinds of questions that I would ask you, it would be our desire to direct the discussion to the level of the undergraduate college student in psychology. Examples of the areas of discussion that would be of interest at this level would be the unconscious, introversion-extroversion and the ways in which these tendencies interact with the factors in your tetrasomy (feeling, thinking, intuition, sensation), the Word-Association Method, views of human personality development and maturity, and so on. Naturally, we shall endeavor in every way to direct our interviews to meet with your complete approval.

On behalf of Dr. Meaney, our psychology department staff, and the University of Houston administration, I wish to thank you for your graciousness in accepting our proposal, thereby allowing our project to begin on such a distinguished note.

Cordially yours,

April 1957

Richard I. Evans, Ph. D. Esq.
University of Houston
Cullen Boulevard
Houston 4, Texas
USA

Dear Prof. Evans,

Thank you for your kind letter. The proposed honorarium of five hundred dollars will suit me completely.

Thank you also for giving me an outline of the questions you are going to ask. I sincerely hope that I shall not be too complicated.

Looking forward to our meeting, I remain, Dear Prof. Evans,

Yours sincerely,

May 16, 1957

Professor Doctor C. G. Jung
Seestrasse 228
Küsnacht-Zurich, Switzerland

Dear Professor Jung:

We were delighted that the honorarium of five hundred dollars will be satisfactory. We are also pleased that the general discussion areas which we listed will be agreeable.

May I raise an additional point? Dr. Meaney, who will, of course, be filming our interviews would like your opinion of the lighting situation which for film work is very important, as you know. For example, at four in the afternoon during the early days of August when we have our appointments with you, is the lighting outside sufficient so that we may actually film the interviews outside in the front of your house, perhaps? From a technical point of view, this would then make it unnecessary to set up special lighting which might be necessary if the interviews were filmed, perhaps, some place in your home. Sound could also be more effectively recorded outside. Your comments concerning these points will be greatly appreciated.

Incidentally, as a matter of routine, our University requires your signature on the enclosed form. We would appreciate it if you would sign it above your name. The extra copy is for your files. Please return the copy bearing your signature with your reply to this letter.

Needless to say, Dr. Meaney and I are very excited about our trip and the prospects of meeting and spending a few hours with you. Our students are already asking us when our filmed reports of the interviews will be available for them to see. Thank you again for making this venture possible for us.

Cordially yours,

May 30, 1957

Professor Richard I. Evans, Ph. D.
University of Houston
Cullen Blvd.
Houston 4, Texas
USA

Dear Dr. Evans,

I assume that you know the chaotic conditions of European weather. Concedente Deo, we have the most beautiful bright sunlight. But, if the Nephele-geretes Zeus prefers to envelop our beloved country in shrouds of mist and rain, it may even happen that we have to put on the lights in the room. If the weather is good and hot, we have a lot of noise near the house on account of a public bathing place. In that case, we should retire to a remote corner of the garden, where there is no electricity. In this case, you would need about 100 yards of wire. Well, I have to leave these technical decisions to yourself.

Here enclosed you also find the signed declaration.

Au revoir in summer!

Sincerely yours,

Like many American psychologists, I had long respected and been interested in Jung as an important pioneer and historical figure in the history of psychology. It was equally true, however, at that time, that I shared some of the skepticism, abundantly in evidence, that surrounded many of Jung's fundamental notions. On the whole, American psychologists found Jung's work too mystical and philosophical to satisfy their criteria for sound, scientific research. In fact, it is interesting to note that at that time, Jung's ideas were more characteristically heralded by members of philosophy and English departments in universities than by the inhabitants of psychology departments.

Contemporary psychology, during the late 1950s, with its emphasis on scientific analysis and procedure, could hardly have been expected to embrace such metaphysical concepts, postulated by Jung, as a "race unconscious" or "transcendental conceptions of the self," much less his suggestion that ancient writings in alchemy can supply knowledge of the growth and development process of the individual.

Few denied, however, that certain of Jung's ideas became conspicuous contributions in psychology. His introvert-extrovert typology had become part of the active vocabulary of countless numbers who had no formal training in psychology; and such classical terms as "complex," only one of many such terms introduced by Jung, had been so well assimilated into our modern language as to make them, in essence, household words. Furthermore, as the creator of the "word association" test, he had supplied a tool which most psychologists found extremely useful.

Also, it would be unfair not to mention that there were a few American psychologists who took Jung's concepts concerning the nature of human personality more seriously. One notable example was the dis-

tinguished, respected, and provocative Dr. Henry
Murray of Harvard University. Dr. Murray, who had
considerable contact with Jung, continues to speak
highly of him. Also, Drs. Calvin Hall and Gardner
Lindzey (1970) in a widely used textbook, *Theories
of Personality*, had written a chapter on Jung that
appeared to me to be one of the most laudatory and
positive descriptions of Jung's ideas to be presented
in the literature of contemporary American psychology. Thus Jung was not entirely without his distinguished following, even in this land of skeptics.

As I approached the idea of interviewing Jung, I
was compelled to decide what the purpose of the
impending interviews should be. Like so many professors in personality courses, I had been teaching
Jungian theory for many years; and, of course, the
teaching situation implies a context where critical
evaluation is very important. In these interviews, however, it seemed best not to create an atmosphere where
critical assessment of Jung's work played any crucial
role.

In order to present Jung's views objectively, I
felt it would be best to provide an opportunity for him
to set forth his views, particularly those views per-
taining to elementary psychology, in as direct and
systematic a fashion as possible. The questions that
I framed would allow him to contrast his views with
those of Freud, and expound upon his own unique
contributions. We used Jung's reactions to Freudian
theory not only as a means of allowing the student to
compare the two men, but also of generating a development of Jung's ideas. We hoped this manner of

presentation, within a framework intelligible to contemporary psychology students, would provide an opportunity for greater clarity in communication.

After arriving in Küsnacht, Switzerland, on August 2, 1957, we immediately contacted Aniela Jaffé, Jung's secretary. I was delighted to hear from her that Dr. Jung would be glad to receive me in his garden the next day to talk very briefly about the interviews that were to take place on the following days. In the meantime, Dr. Meaney had the tremendous task of setting up the film equipment in the Swiss Federal Institute of Technology, the site of the forthcoming interviews. We had originally planned to film the interviews in Jung's garden, but technical problems with our equipment and available electrical current made a change necessary.

The problems of our one-man camera operator, director, and coproducer, Dr. Meaney, to get a technical crew together became quite a chore. As he busied himself with these problems the next day, I walked over from our hotel to Dr. Jung's home a few blocks away. His first words to me were, "Why do you American psychologists hate me so much?" Even on such short acquaintance, I did not react to this disarming statement as seriously as might be expected, because of the teasing twinkle I detected in Dr. Jung's eyes.

Although he was at that time eighty-two years old, except for some marked deafness, he appeared to be in excellent health and physically was a well-proportioned figure of a man, over six feet tall, who carried himself with dignity. His manner was warm and charming, and it was easy to gain rapport with him. I might add, however, that his façade was quite deceptive. One felt a probing, analytic stance underneath the charm. Knowing his distrust of Americans in general, and of American psychologists in particular, it was difficult to assess his true attitude toward me, and toward the prospects of the interviews.

I answered Dr. Jung's initial question with the only legitimate response available; that is, I had to admit that there were certainly a large number of American psychologists, probably the majority of them in fact, who did not accept many of his ideas, although "hate" seemed a rather a strong term to use in reference to their sentiments. I pointed out to him, however, that another group of psychologists existed in America who were quite familiar with his work and much more positive in their evaluation of it.

Ensuing discussions revealed that he was indeed aware of the critical scrutiny to which certain of his concepts had been subjected. For example, it was no secret to him that his introversion-extroversion typology had served many of our introductory textbooks in psychology well as a kind of whipping boy in an attempt to warn the beginning student not to type people or put them into categories; in this case to label them as simply introverts or extroverts. On this point, however, I explained to him that many psychologists, perhaps those more familiar with his work, were cognizant that he had never intended these typologies to be anything more than a useful guide in helping to understand the individual.

Tea was served, and we began discussing a number of different topics. Dr. Jung was particularly interested in the educational goals of our project and wanted to know for whom these recorded discussions were intended. I expressed a hope that they would be a vehicle for introducing students to his work, which made it necessary that they be couched in a level of language comprehensible to even the beginning student. He indicated a clear understanding of this.

As we talked the educational nature of our effort appeared to evoke genuine interest from him, and I asked him why he had agreed to participate in these interviews. He replied that he somehow intuitively felt that these professors from Houston, The University of Houston, in "Houston, Texas, a new frontier in the United States," would be anxious to handle this situation in a way of which he would approve.

Jung's English, couched in a distinct German accent, was quite good and I began to see a dynamic quality in his mannerisms, his expressiveness, and his colorful phrasing. As we talked, it became very evident that in Carl Jung we had not only a subject who had much to offer intellectually, but an individual who would give an excellent account of himself in a spontaneous interview situation.

I suggested to Dr. Jung that if true spontaneity could be achieved in these interviews, it would be much more exciting and interesting to the student, to which he immediately acknowledged his agreement. As a matter of fact, although I had prepared the questions to be asked during the four one-hour interviews and sent him copies, he did not seem to be concerned about discussing them after he read them.

The stress on spontaneity as a highly desirable goal, coupled with the one-hour time span allotted for each interview, however, created an important consideration to be taken into account. Spontaneity can result in too much irrelevant and digressive conversation, and in this case, since we wanted Jung's reactions to all the prepared questions, every hour had to count. I pointed this out to him, explaining that in commercial efforts of this type, with more interview time available, many hours are often edited down to a few hours in the completed film or publication. In these interviews, we would have virtually no leeway for editing. Again, Jung indicated that he understood.

In the course of completing the four one-hour interviews, a number of interesting things happened that afforded us further insight into Jung as a person; although it should be kept in mind that these limited contacts could hardly be expected to supply a base for a truly profound analysis of Jung at eighty-two.

Each day, in a rented Plymouth, we proceeded to the site of the interviews, the Swiss Federal Institute of Technology. I had carefully mapped out the way so that I would encounter no difficulty in locating the Institute. This proved an unnecessary precaution. I discovered that Jung was fond of driving all over Europe himself, and knew every corner of this city. Each day he would point the way, always showing me a new route, proclaiming, "Everyone knows that Jung never goes the same way twice." Arriving at the Institute, I attempted to lead him to an elevator for the trip to the second floor, but he boldly led the way up the stairs, tiring me with his brisk pace.

Time magazine had assigned Tom Dozier to cover the story of these interviews, and Jung and I both invited him to sit in on the interviews, which he followed closely. On one occasion, he rode back from the Institute with us, sitting in the back seat while Jung and I conversed in an informal and casual manner about Jung's grandchildren and great-grandchildren. Dozier interrupted our chat to ask Jung a really excellent question about some highly technical point discussed in that day's interview. Jung declared, "Sir, we are at the moment talking about something a little bit more important. Why don't you ask Dr. Evans later?"

Time had previously published stories favorable to Jung, so his behavior in this case was not a reflection of hostility toward the magazine or Tom Dozier, but rather demonstrated Jung's lack of concern about making an impression on the media. Popular culture certainly did not appear to be significant to him.

Dozier's description of our efforts appeared in the August 19, 1957, edition of *Time*. An abstract from his story, prior to editing, appeared in the Houston *Post*, September 16, 1957:

> The old man with the wispy white hair and the wisely twinkling eyes leaned back in an armchair, deliberately puffing his pipe. Seemingly unconscious of the microphone around his neck and the camera lens staring across the room at him, Carl Gustav Jung spoke through the smoke wreathing his head. His voice was strong and booming, his English only slightly tinged with a German accent.
>
> "The world," said Jung, "hangs on a thin thread,

and that thread is the psyche of man. . . . It is not the reality of the hydrogen bomb we must fear, but what man does with it. Suppose certain fellows in Moscow lose their nerve, then the world is in fire and flames. As never before, the world hangs on the psyche of man."

Therefore, explained the wise old man, the study and understanding of man's psyche is more important than ever.

For an hour a day on four different days, Analytical Psychologist Jung, at a virile 82, the last survivor among the Big Three founding fathers of modern psychology, sat before a television camera in a glass-walled room in Zurich's Federal Institute of Technology, and explained the abstruse fine points of the Jungian approach to the study of man's mind. Guided gently by Interviewer Richard Evans of the University of Houston's Department of Psychology, Jung ranged over the whole voluminous complexity of his theories and conclusions about the psyche.

At times he flailed out mildly at his fellow Titans, Freud and Adler, repeatedly corrected what he considered misinterpretations of his ideas, explained in detail his theories on introversion, extroversion, the persona, intuition, the interpretation of dreams and the unconscious symbols called archetypes . . .

Jung's performance was as rare as it was fascinating. He was appearing for the first time ever before a TV camera, was making his first bow before an American audience since he lectured on psychology and religion at Yale in 1938, and except for lectures in Zurich, his only public declamation in a decade. And despite his hearty good health, onlookers were impressed by the possibility that they might be witnessing the last curtain speech of a great and masterful actor . . . a retiring genius who shuns public appearances because "I have such a hell of a trouble to make people see what I mean.". . .

. . . Jung quipped and sparkled, seemed to enjoy the whole thing immensely. As Meaney fastened the microphone around Jung's neck and attached the lead-in wire, the old man joked, "Well, this is the first time anybody ever had me on a leash.". . . At the studio, his eyes sparkled behind steel-rimmed spectacles, and his bristly white mustache wiggled when he smiled. Since he was lecturing for students and since he is often obscure anyway, much of what he said was highly technical and difficult to translate into everyday English. . . .

By the fourth day, Jung appeared to tire somewhat. We were almost ready not to press for completion of the last interview, but he insisted that he wanted to complete it. In fact, he cheerfully indicated that he was really enjoying it. I think that the process of educating a great number of students in this manner represented a genuine challenge to him.

Later we sent him a copy of one of the filmed interviews as a gift. We received the following letter from his secretary, Aniela Jaffé:

November 28, 1958

Prof. Richard I. Evans
University of Houston
Cullen Blvd.
Houston 4, Texas

Dear Professor Evans,

I want to tell you how very much we all liked to see your film. It was a great success, and we want to repeat the performance in spring. Prof. Jung asked me to thank you very much indeed for having sent the copy. He himself was not present at this performance, but we hope that he will come the next time. We are sure he will like it.

He asks you whether the film is meant as a gift to him, or whether he has to return it. He should be grateful for a short reply.

We heard that you have four such films. Is that true?

Thanking you again—

Sincerely yours,
(Mrs.) Aniela Jaffé, Secy.

We heard subsequently that he really was quite delighted with the way he appeared in these interviews.

Upon returning to the United States, we developed the films and began considering means by which they could be made available to students and other interested groups. Much to our delight, interest in the films has been very great, and we have filled requests for prints of them from literally all over the world.

As is so often the case with those of us in the field of psychology, we are not only interested in getting a product, but we want to evaluate it very carefully. Without burdening the reader with an elaborate discussion concerning the evaluations we made of these films, citation of one study we did will suffice as an indication, to some extent, of the success of the project. The study compared matched groups of students who read the transcripts of the interviews and saw the filmed interviews, in terms of how much they learned beyond the normal reading of Jung's ideas from other sources. The results of our findings suggested that the interviews not only facilitated more effective communication of Jung's ideas, but also seemed to have definite effect on attitudes and feelings about Jung. Even some of the psychology students who had formerly been extremely inhospitable to Jung's views seemed to develop a more favorable attitude toward the man when confronted with this technique of presenting his ideas. This to us, of course, was most interesting and certainly suggested that the interview technique had definite promise as an educational tool.

In the pages that follow, we have attempted to organize the materials in the interviews in such a way as to facilitate maximum communication between reader and printed page. We have not materially modified Jung's answers to the questions in the first three interviews. In the last interview, however, we actually asked Jung very few questions, allowing him to speak quite freely and in an unguided manner. Somewhat naturally, this approach led to a certain amount of rambling and did not seem to elicit the succinct kind of responses that would have been most ideal. Moreover, he mostly elaborates and expands, in this last interview, his earlier answers to questions posed in the first three; thus, we have taken the liberty of synthesizing and reorganizing, to some extent at least, these expansions upon points already made, in order to render the text more communicative and readable.

The order in which the questions and answers occurred has at times been changed for purposes of clarity, and, in the same interest, some of the questions have been subject to minor revision to make them more succinct. As a whole, however, the material embodied in the succeeding chapters faithfully reflects Jung's responses to our questions.

As the reader begins to delve into this material, he may notice certain differences in the way an American psychologist would use specific terms and in the way these same terms are used by Dr. Jung. To some degree, this is a function of Jung's German as it influences his English. To cite one example, the word "instinct" in American psychology can be defined as an unlearned response tendency. Jung uses the term "instinct" in reference to a learned response in the sense of being a habit.

The reader may also notice that now and then there is a tendency for a phrase to be used, which at face value does not appear particularly meaningful. For example, at one point Jung is talking about the unconscious and indicates that we can't really know much about the unconscious because— ". . . it's really unconscious!" At such times we recall the twinkle in Jung's eyes as he attempted to poke a little humor in the direction of the interviewer, and perhaps to open the door for some further discussion. In other words, the reader should not always take his expressions too literally, because of Jung's sense of humor and his unique communication devices. Many of his responses must be understood in the context of the total answer to a given question.

It was with much interest that I read in the book, *Memories, Dreams, Reflections* (1963) some records of Jung's ideas presented by Aniela Jaffé. Her keen insight, perceptiveness, and sensitive understanding of Jung's ideas and Jung as an individual are deserving of a special note of praise and admiration.

In New York City on December 1, 1961, a Jung Memorial meeting was held, sponsored jointly by the New York Association of Analytical Psychology and the Analytical Psychology Club of New York. At this meeting I was particularly taken with the eloquent comments expressed by Dr. Henry Murray about the late Carl Jung, as he said:

> Jung was humble before the ineffable mystery of each variant self that faced him for the first time, as he sat at his desk, pipe in hand, with every faculty in tune, brooding on the portent of what was being said to him. And he never hesitated to acknowledge his perplexity in the presence of a strange and inscrutable phenomenon, never hesitated to admit the provisional nature of the comments he had to make or to emphasize the difficulties and limitations of possible achievement in the future.
>
> "Whoever come to me," he would say, "takes his life in his hand." The effect of such a statement, the effect of his manner of delivering his avows of uncertainty and suspense is not to diminish but to augment the patient's faith in his position's invincible integrity, as well as to make plain that the patient must take the burden of responsibility for any decisions he might make.

In the text of the discussions that follow, I sincerely hope that the reader will be able to vicariously interact with Jung as well as learn from him.

I. REACTIONS TO FREUD AND VARIOUS PSYCHOANALYTIC CONCEPTS

In this portion of the interviews, Dr. Jung and I probed into the events surrounding his initial involvement with Freud. Also, we attempted to trace the line of the skeletal structure of psychoanalytic theory, allowing Jung to react to each part of it.

As Jung responds to questions concerning psycho-sexual development, and the Freudian concepts of the id, ego, and superego, a surprising degree of insight emerges concerning the manner in which Jung disagreed with Freud, the areas in which they were in concordance, and certain of Jung's ideas which developed as a reaction to Freudian thought.

Freud,
Adler, and
Rank

DR. EVANS: Dr. Jung, many of us who have read a great deal of your work are aware of the fact that in your early work you were to some degree, at least, in association with Dr. Sigmund Freud, and I know it would be of great interest to many of us to hear how you happened to hear of Dr. Freud and how you happened to become involved with some of his work and ideas.

DR. JUNG: Well, as matter of fact, it was in the year 1900, in December, soon after Freud's book about dream interpretation had come out, that I was asked by my chief, Professor Bleuler, to give a review of the book. I studied the book very attentively, and I didn't understand many things in it, which were not at all clear to me. But from other parts I got the impression that this man really knew what he was talking about, and I thought that this is certainly a masterpiece—full of future.

I had no ideas then of my own; I was just beginning. It was just when I began my career as assistant in the psychiatric clinic. And then I began with experimental psychology or psychopathology. I applied the experimental association methods of Wundt, the same that have been applied in the psychiatric clinic in Munich, and I studied the results and had the idea that one should go once more over it, so I made these association tests, and I found out that the important thing in them has been missed, because it is not interesting to see that there is a reaction—a certain reaction—to a stimulus word; that is more or less uninteresting. But the interesting thing is why people could *not* react to certain stimulus words, or only react in an entirely inadequate way.

Then I began to study these places in the experiment where the attention, or the capability of the test person apparently began to waver or to disappear, and I soon found out that if it was a matter of intimate personal affairs people were thinking of, or which were in them, even if they momentarily did not think of them, when they were unconscious in other words; that the inhibition came from the unconscious and hindered the expression in speech. Then, in examining all these cases as carefully as possible, I saw that it was a matter of what Freud called repression. I also saw what he meant by symbolization.

DR. EVANS: In other words, from your word association studies, some of the things in *The Interpretation of Dreams* (Freud, 1950) began to fall into place?

DR. JUNG: Yes! And then I wrote a book about the psychology of dementia praecox, as it was called then —now it is schizophrenia—and I sent the book to Freud, writing to him about my association experiments and how they confirmed his theory thus far. That is how my friendship with Freud began.

DR. EVANS: There were other individuals who also became interested in Dr. Freud's work, and one of them was Dr. Alfred Adler. As you remember Dr. Adler, what in your estimation led him to become interested in Dr. Freud's work?

DR. JUNG: He belonged; he was one of the young doctors that belonged to his surroundings there. There were about twenty young doctors who followed Freud there, who were . . . who had a sort of little society. Adler was one who happened to be there, and he learned—he studied Freud's psychology in that circle.

DR. EVANS: Another individual, of course, who joined this group was Otto Rank, and he, unlike yourself, Dr. Adler, and Dr. Freud, was not a physician; did not have the Doctor of Medicine degree. Was this regarded by your group at the time as something unusual, to have someone become interested in these ideas who was not by training a physician?

DR. JUNG: Oh no! I have met many people who represented different faculties who were interested in psychology. All people who had to do with human beings were naturally interested; theologians, lawyers, pedagogues; they all have to do with the human mind and these people were naturally interested.

DR. EVANS: Then your group, including Freud, did not feel that this was exclusively an area of interest for the physician? This was something that might appeal to many?

DR. JUNG: Oh my, yes! Mind you, every patient you have gets interested in psychology. Nearly everyone thinks he is meant to be an analyst, inevitably.

Psychosexual Development

DR. EVANS: One of the very fundamental ideas of the original psychoanalytic theory was Freud's conception of the libido as a sort of broad, psychic sexual energy. Of course, we all know that you began to feel that Dr. Freud might have laid, perhaps, a little too much stress on sexuality in his theories. When did you first begin feeling this?

DR. JUNG: In the beginning, I had naturally certain prejudices against these conceptions and after a while I overcame them. I could do that owing to my biological training. I could not deny the importance of the sexual instinct, you know.

45

Later on, however, I saw that it was really one-sided because, you see, man is not only governed by the sex instinct; there are other instincts as well. For instance, in biology you see that the nutritional instinct is just as important as the sex instinct, although in primitive societies sexuality plays a role much smaller than food. Food is the all-important interest and desire. Sex—that is something they can have everywhere—they are not shy. But food is difficult to obtain, you see, and so it is the main interest.

Then in other societies—I mean civilized societies —the power drive plays a much greater role than sex. For instance, there are many big businessmen who are impotent because their full energy is going into moneymaking or dictating the laws to everybody else. That is much more interesting to them than affairs with women.

DR. EVANS: So in a sense, as you began to look over Dr. Freud's emphasis on sexual drive, you began to think in terms of other cultures, and it seemed to you that this emphasis was not of sufficient universality to be assessed primary importance.

DR. JUNG: Well, you know, I couldn't help seeing it, because I had studied Nietzsche. I knew the work of Nietzsche very well. He had been a professor at Basel University, the air was full of talk about Nietzsche so naturally I had studied his works. And there I saw an entirely different psychology, which was also a psychology—a perfectly competent psychology—but all built upon the power drive.

DR. EVANS: Do you think it possible that Dr. Freud

was either ignoring Nietzsche, or had perhaps not
wanted to be influenced by Nietzsche?

DR. JUNG: You mean his personal motivation?

DR. EVANS: Yes.

DR. JUNG: Of course it was a personal prejudice.
It happened to be his main point, you know, that
certain people are chiefly looking for this side, and
other people are looking for another side. So, you see,
the inferior Dr. Adler, the younger, the weaker, naturally had a power complex. He wanted to be *the*
successful man. Freud was a successful man; he was
on top, and so he was interested only in pleasure and
the pleasure principle, and Adler was interested in the
power drive.

DR. EVANS: You feel that it was a sort of function
of Dr. Freud's own personality?

DR. JUNG: Yes, it is quite natural; it is one of two
ways how to deal with reality. Either you make reality
an object of pleasure, if you are powerful enough
already; or you make it an object of your desire to
grab it or possess it.

DR. EVANS: Some observers have speculated that
the patients whom Dr. Freud saw in the Vienna of this
period were so often sexually repressed that they may
have been representative of a cultural type; or, in
other words, since these patients were a part of a
Viennese society, believed to have been a rather
repressed society, Freud's patients, perhaps, demonstrated an undue tendency to react to sexual frustrations, reinforcing his ideas of a sexual libido.

DR. JUNG: Well, it is certainly so that at the end

of the Victorian age there was a reaction going on over the whole world against the sex taboos, so-called. One didn't properly understand anymore why or why not; and Freud belongs in that time, a sort of liberation of the mind of such taboos.

DR. EVANS: There was a reaction, then, against the sort of tight, inhibited culture he was living in.

DR. JUNG: Yes, Freud, in that way—on that side— really belonged to the category of a Nietzschean mind. Nietzsche had liberated Europe from a great deal of such prejudices, but only concerning the power drive and our illusions as to the motivations of our morality. It was a time critical of morality.

DR. EVANS: So Dr. Freud, in a sense, was taking another direction.

DR. JUNG: Yes. And then, moreover, sex being the main instinct—the predominating instinct in a more or less safe society, when the social conditions are more or less safe—sexuality is apt to predominate because people are taken care of. They have their positions. They have enough food. When there is no question of hunting and seeking food, or something like that, then it is quite probable that patients you meet more or less all have a certain sexual complex.

DR. EVANS: So the sex drive is potentially the drive in that particular society most likely to be inhibited?

DR. JUNG: Yes. It is a sort of finesse, almost, when you find out that somebody has a power drive and their sex only serves the purpose of power. For instance, a charming man whom all women think is the real hero of all hearts; he is a power-devil, like a Don Juan, you know. The woman is not his problem; his problem is how to dominate. So in the second place after sex comes the power drive. And even that is not the end.

DR. EVANS: To proceed further, in the orthodox psychoanalytic view, as you well know, there is much attention paid to what Freud called psychosexual development, in that the individual encounters a series of problems in sequence which he must resolve in order to progressively mature. It appears that one of the earliest problems the individual seems to have centers around, you might say, primitive oral satisfactions, or oral zone experiences—including weaning —representing some of the first frustrations for the infant.

DR. JUNG: I think, you see, that when Freud says that one of the first interests, and the foremost interest is to feed, he doesn't need such a peculiar kind of terminology like "oral zone." Of course, they put it into the mouth . . .

DR. EVANS: Then you look at Freud's oral level of development in a less complicated sense *without* a sexual connotation?

DR. JUNG: Science consists to a great extent of mere talk.

DR. EVANS: In summary then, Dr. Jung, with reference to the oral level of development, you prefer to look at it rather literally, as a sort of hunger drive or drive for nutrition.

Another rather fundamental point in the development of the ego in the orthodox psychoanalytic view is that the oral level is followed by another critical level, an anal level of development. At this level another crucial, early frustration arises; that is, frustration centering around the problem of toilet-training. In ego development and later character formation, Freud saw poor resolution of such problems as being rather serious.

DR. JUNG: Well, one can use such terminology because it is a fact that children are exceedingly interested in all orifices of the body and in doing all sorts of disgusting things, and sometimes such a peculiarity keeps on into later life. It is quite astonishing what you can hear in this respect. Now it is equally true that people who have such preferences also develop a peculiar character.

In early childhood a character is already there. You see, a child is not born *tabula rasa* as one assumes. The child is born as a high complexity, with existing determinants that never waver through the whole life, and that give the child his character. Already, in earliest childhood, a mother recognizes the individuality of her child; and so, if you observe carefully, you see a tremendous difference, even in very small children.

These peculiarities express themselves in every way. First, the peculiarities express themselves in all childish activities—in the way it plays, in the things it is interested in. There are children who are tremendously interested in all moving things, in the movement chiefly, and in all things they see that affect the body. So they are interested in what the eyes do, what the ears do, how far you can bore into the nose with your finger, you know. They will do the same with the anus; they will do whatever they please with their genitals. For instance, when I was in school, we once stole the class book where all the punishments were noted, and there our professor of religion had noted, "So-and-so punished with two hours because he was toying with his genitals during the religious hour."

These interests express themselves in a typically childish way in children. Later on they express themselves in other peculiarities which are still the same, but it doesn't come from the fact that they once had done such and such a thing in childhood. It is the character that is doing it. There is a definite inherited complexity, and if you want to know something about possible reasons, you must go to the parents.

So in any case of a child's neurosis, I go back to the parents and see what is going on there, because children have no psychology of their own, literally taken. They are so much in the mental atmosphere of the parents, so much a participation mystique with the parents. They are imbued by the maternal or paternal atmosphere, and they express these influences in their childish way. For instance, take an illegitimate child. They are particularly exposed to environmental difficulties such as the misfortune of the mother, etc., etc., and all the complications. Such a child will miss, for instance, a father. Now in order to compensate for this, it is just as if they were choosing or nominating a part of their body for a father, a substitution for the father, and they develop, for instance, masturbation. That is very often so with illegitimate children; they become terribly autoerotic, even criminal.

DR. EVANS: With reference to the role of the parents in development, one of the central parts of psychosexual development in orthodox psychoanalytic theory is the Oedipal level of development. It is at this level that the problem of premature sexuality relating to the opposite-sex parent emerges. This problem, like the earlier ones mentioned, must also be resolved, or it will result in the formation of an Oedipus complex.

DR. JUNG: That is just what I call an archetype. That is the first archetype Freud discovered; the first and the only one. He thought that this *was* the archetype. Of course, there are many such archetypes. You look at Greek mythology and you find them, any amount of them. Or look at dreams and you find any amount of them. To Freud, however, incest was so impressive that he chose the term "Oedipus complex," because that was one of the outstanding examples of an incest complex; though, mind you, it is only in the masculine form, because women have an incest complex too which, to Freud, was not an Oedipus. So it is something else.

So it is only a term for an archetypal way of behavior, in the case of a man's relation, say, to his mother. But it also means to his daughter because whatever he was to the mother, he will be it to the daughter too. It can be this way or that way. It depends.

DR. EVANS: Then you believe, in other words, that the Oedipus complex is not as important an influence in itself as Freud did, but that it is only one of many archetypes?

DR. JUNG: Yes. It is only one of the many, many ways of behavior. Oedipus gives you an excellent example of the behavior of an archetype. It is always a whole situation. There is a mother, there is a father, there is a son; so there is a whole story of how such a situation develops and to what end it leads finally. That is an archetype.

An archetype always is a sort of abbreviated drama. It begins in such and such a way, it extends to such and such a complication, and finds its solution in such and such a way. That is the usual form. For instance, take the instinct in birds of building their nests. In the way they build the nest, there is the beginning, the middle, and the end. The nest is built just to suffice for a certain number of young. The end is already anticipated. That is the reason why, in the archetype itself, there is no time; it is a timeless condition where beginning, middle, and end are just the same; they are all given in one. That is only a hint to what the archetype can do, you know. But that is a complicated question.

DR. EVANS: To discuss more specifically Freud's concept of the Oedipus complex, a rather commonly held belief, again within the confines of orthodox psychoanalytic theory, is that, in a sense, the child's early family behavior patterns with the mother, the father, etc., are to some extent repeatedly relived and can be regarded as a "repetition compulsion." For example, when the young man gets married, he may to some degree react to his wife as he did to his mother, or he may be searching for someone like his mother. Likewise, the daughter, as she looks for a husband, may be searching for a father. This will be repeated over and over again.

This appears to be the heart of what the early Freudians were theorizing. Now, does this type of recapitulation of the very early Oedipus situation fit in with your conceptions?

DR. JUNG: No. You see, Freud speaks of the incest complex just in the way you describe, but he omits completely the fact that with this Oedipus complex, he is only giving the contrary—namely, the resistance against it. For instance, if the Oedipus pattern were really predominant, we would have been suffocated in incest half a million years ago, at least.

But there is a compensation. In all of the early levels of civilization you find the marriage laws— namely, exogamic laws. The first form, the most elementary form, is that the man can only marry his cousin on the maternal side. The next form is that the man can only marry his cousin in the second degree, namely, from the grandmother. There are four systems; quarternary systems and systems of eight and twelve, also a six-system. In China there are still traces of both a twelve-system and of a six-system. And those are developments beyond the incest complex and against the incest complex. Now if sexuality is predominant—particularly incestual sexuality—how can it develop?

These things have developed in a time long before there was an idea of the child, say, of my sister. That's all wrong. On the contrary, it was a royal prerogative as late as the Achaemenite kings of Persia, and among the Egyptian Pharaohs. If the Pharaoh had a daughter from his sister, he married that daughter and had a child with her, and then married his granddaughter. Because that was royal prerogative. You see, the preservation of the royal blood is always a sort of attempt at the highly appreciated incestuous restriction of the numbers of ancestors, because this is loss of ancestors. Now, you see, that must be explained too. It is not only the one thing but is also its compensation. You know this plays a very great role in the history of human civilization.

Freud is always inclined to explain these things by external influences. For instance, you would not feel hampered in any way if there were not a law against it. No one is hampered by one's self. And that's what he never could admit to me.

"Id," "Ego," and "Superego"

DR. EVANS: Going still further into the development of Dr. Freud's theory, which you acknowledge as a significant factor in the development of many of your own early ideas, Dr. Freud, of course, talked a great deal about the unconscious.

DR. JUNG: As soon as research comes to a question of the unconscious, things become necessarily blurred, because the unconscious is something which is really unconscious! So you have no object—nothing. You can only make inferences because you can't see it, and so you have to create a model of this possible structure of the unconscious.

56

Now Freud came to the concept of the unconscious chiefly on the basis of the same experience I have had in the association experiment; namely, that people reacted—they said things, they did things—without knowing that they had done it or had said it. This is something you can observe in the association experiment; sometimes people cannot remember afterward what they did or what they said in a moment when a stimulus word hits the complex. In the so-called reproduction experiment, you go through the whole list of words. You see that the memory fails when there is a complex reaction or block. That is the simple fact upon which Freud based his idea of the unconscious.

There is no end of stories, you know, about how people can betray themselves by saying something they didn't mean to say at all; yet the unconscious meant them to say just that thing. That is what we can see, time and again, when people make a mistake in speech or they say something which they didn't mean to say; they just make ridiculous mistakes. For instance, when you want to express your sympathy at a funeral, you go to someone and you say, "I congratulate you." That's pretty painful, you know, but that happens, and it is true. This is something that goes parallel with the whole school of the Salpetrière, Paris.

There was Pierre Janet who worked out another side of the understanding of unconscious reactions. Now, Freud refers very little to Pierre Janet, but I studied with him in Paris and he very much helped form my ideas. He was a first-class observer, though he had no systematic, dynamic psychological theory; it was a sort of physiological theory of the unconscious phenomena.

There is a certain depotentiation of the tension of consciousness; it sinks below the level of consciousness and thus becomes unconscious. That is Freud's view too, but he says it sinks down because it is helped; it is repressed from above. That was my first point of difference with Freud. I think there are cases in my observations where there was no repression from above; those contents that became unconscious had withdrawn all by themselves, and not because they were repressed.

On the contrary, they have a certain autonomy. Then I discovered the concept of autonomy in that these contents that disappear have the power to move independently of my will. Either they appear when I want to say something definite; they interfere and speak themselves instead of helping me to say what I want to say; they make me do something which I don't want to do at all; or they withdraw in the moment that I want to use them. They certainly disappear!

DR. EVANS: And this then is independent of any of the, you might say, pressures on the consciousness as Freud suggested?

DR. JUNG: Yes. There can be such cases, sure enough, but besides them, there are also the cases that show the unconscious contents acquire a certain independence. All mental contents having a certain feeling tone that is emotional have the value of an emotional affect—have the tendency to become autonomous. So, you see, anybody in an emotion will say and do things which he cannot vouch for. He must excuse himself that he was in a state; he was *non compos mentis.*

DR. EVANS: Dr. Freud suggested that the individual is born under the influence of what he called the id, which is unconscious and undeveloped, a collection of animal drives. It is not very easily understood where all these primitive drives—all these instincts—come from.

DR. JUNG: Nobody knows where instincts come from. They are there and you find them. It is a story that was played millions of years ago. There sexuality was invented, and I don't know how this happened; I wasn't there! Feeding was invented very much longer ago than even sex, and how and why it was invented, I don't know. So we don't know where the instinct comes from. It is quite ridiculous, you know, to speculate about such an impossibility. So the question is only: where do those cases come from where instinct does not function? That is something within our reach, because we can study the cases where instinct does not function.

DR. EVANS: Could you give us some rather specific examples of what you mean by cases where the instinct does not function?

DR. JUNG: Well, you see, instead of instinct, which is a habitual form of activity, take any other form of habitual activity. Suddenly a thing doesn't function. Take a singer who is absolutely controlling his voice, suddenly he can't sing. Or take a man who writes fluently, suddenly he makes a ridiculous mistake; there his habit doesn't function. You see, when you ask me something, I'm supposed to be able to react to you; suddenly if I am *bouche béante* [open-mouthed], or if you succeed in touching upon one of my complexes, you will see that I become absolutely perplexed; I am *depossedé*, words fail me.

DR. EVANS: We haven't seen you very perplexed yet, Dr. Jung.

DR. JUNG: Or look at exam psychology, you know, a fellow who knows his stuff quite well—the professor asks him and he cannot say a word.

DR. EVANS: To continue, another part of Dr. Freud's theory, of course, that became very important, to which we have already alluded, was the idea of the conscious; that is, out of this unconscious, instinctual structure, the id, an ego emerges. Freud suggested that this ego resulted from the organism's contact with reality, perhaps a product of frustration as reality is imposed on the individual. Do you accept this concept of the ego?

DR. JUNG: That man has an ego at all, that is your question? Ah, that is again such a case as before; I wasn't there when it was invented. However, in this case, you see, you can observe it to a certain extent with a child. A child definitely begins in a state where there is no ego, and about the fourth year or before, the child develops a sense of ego—"I, myself."

There is, in the first place, a certain identity with the body. For instance, when you ask primitives, they emphasize always the body. When you ask, "Who has brought this thing here," the Negro will say "I brought it." No accent on the "I," simply "brought

it." Then if you say, "But have YOU brought it," he
will say, "in here, ME, ME, Yes, I, MYSELF"—this
given object, this thing here. So the identity with the
body is one of the first things which makes an ego; it
is the spatial separateness that induces, apparently,
the concept of an ego.

Then, of course, there are lots of other things.
Later on there are mental differences and personal
differences of all sorts. You see, the ego is continuously
building up; it is not a finished product—never—it
builds up. You see, no year passes when you do not
discover a new little aspect in which you are more
ego than you have thought.

DR. EVANS: Dr. Jung, there has been much discussion about how certain experiences in the early
years influence the formation of the ego. For example,
one of the most extreme views concerning such early
influences was advanced by Otto Rank. He spoke of
the birth trauma and suggested that the trauma of
being born would not only leave a very powerful impact on the developing ego, but would have residual
influence throughout the life of the individual.

DR. JUNG: I should say that it is very important
for an ego that it is born; this is highly traumatic,
you know, when you fall out of heaven.

DR. EVANS: However, do you take literally Otto
Rank's position that the birth trauma has a profound
psychological effect on the individual?

DR. JUNG: Of course, for instance, if you are a
believer in Schopenhauer's philosophy, you say, "It
is a hellish trauma to be born." Well, there is a Greek
saying: "It is beautiful to die in youth, but the most

beautiful of all things is not to be born." Philosophy, you see.

DR. EVANS: But you don't take this as a literal psychic event?

DR. JUNG: Don't you see, this is an event that happens to everybody that exists—that each man once has been born. Everybody who is born has undergone that trauma, so the word trauma has lost its meaning. It is a general fact, and you cannot say, "It is a trauma." It is just a fact, because you cannot observe a psychology that hasn't been born—only then you could say what the birth trauma is. Until then, you cannot even speak of such a thing; it is just a lack of epistemology.

DR. EVANS: In his later writing, in addition to the ego, Freud introduced a term to describe a particular function of the ego. That term was the superego. Broadly speaking, the superego was to account for the moral restrictive function of the ego.

DR. JUNG: Yes, that is the superego, namely that codex of what you can do and what you cannot do.

DR. EVANS: Built-in prohibitions which Freud thought might be partly acquired and partly "built-in?"

DR. JUNG: Yes. However, Freud doesn't see the difference between the "built-in" and the acquired. You see, he must have it almost entirely within himself; otherwise, there could be no balance in the individual. And who in hell would have invented the Decalogue? That is not invented by Moses, but that is the eternal truth in man himself, because he checks himself.

II. THE UNCONSCIOUS

Perhaps the area of greatest concentration and analysis in Jung's theory is that of the unconscious. In contrast to Freud's development of a single unconscious which, particularly in his earlier work, was the source of all pleasure-seeking, instinctual urges within the individual and the domain of repressed material as well, Jung postulated both a personal unconscious and a race or group unconscious, perhaps his most controversial contribution. Of particular importance in the race unconscious is Jung's contention that archetypes—innate behavior potentials—inherited in what might best be described as a quasi-Lamarckian sense, are the crucial determiners of human development.

In these interviews, Jung explicitly explains archetypes and related concepts such as Persona, the Ego, and the Self. It becomes clear that Freud's concept, the Ego, to him the unifying core of human personality, is essentially Jung's concept of the Self.

Archetypes

DR. EVANS: You mentioned earlier that Freud's Oedipal situation was an example of an archetype. At this time would you please elaborate on the concept, archetype?

DR. JUNG: Well, you know what a behavior pattern is, the way in which, say, a weaverbird builds its nest. That is an inherited form in him which he will apply. Or certain symbiotic phenomena between insects and plants are inherited patterns of behavior. And so man has, of course, an inherited scheme of functioning. You see, his liver, his heart, all his organs, and his brain will always function in a certain way, following its pattern. You and I have a great difficulty seeing it because we cannot compare. There are no other similar beings like man, that are articulate, that could give an account of their functioning. If that were the case, we could . . . I don't know what. But because we have no means of comparison, we are necessarily unconscious about our own conditions.

It is quite certain, however, that man is born with a certain functioning, a certain way of functioning, a certain pattern of behavior which is expressed in the form of archetypal images, or archetypal forms. For instance, the way in which a man should behave is expressed by an archetype. Therefore, you see, the primitives tell such stories. A great deal of education goes through storytelling. For instance, they call in a palaver of the young men and two older men perform before the eyes of the younger all the things they should not do. Then they say, "Now that's exactly the thing you shall not do." Another way is they tell them of all the things they should not do, like the Decalogue —"Thou shalt not"—and that is always supported by mythological tales.

That, of course, gave me a motive to study the archetypes, because I began to see that the structure of what I then called the "collective unconscious" was really a sort of agglomeration of such typical images, each of which had a numinous quality.

The archetypes are, at the same time, dynamic. They are instinctual images that are not intellectually invented. They are always there and they produce certain processes in the unconscious that one could best compare with myths. That's the origin of mythology. Mythology is a pronouncing of a series of images that formulate the life of archetypes.

So the statements of every religion, of many poets, etc., are statements about the inner mythological process, which is a necessity because man is not complete if he is not conscious of that aspect of things. For instance, our ancestors have done so-and-so, and so shall you do. Or such and such a hero has done so-and-so, and that is your model. For instance, in the teachings of the Catholic Church, there are several thousand saints. They serve as models, they have their legends, and that is Christian mythology.

In Greece, you know, there was Theseus and there was Heracles, models of fine men, of gentlemen, you know; and they teach us how to behave. They are archetypes of behavior. I became more and more respectful of archetypes, and that naturally led me on to a profound study of them. And now, by Jove, that is an enormous factor, very important for our further development and for our well-being, that should be taken into account.

It was, of course, difficult to know where to begin, because it is such an enormously extended field. And the next question I asked myself was, "Now, where in the world has anybody been busy with that problem?" I found that nobody had except a peculiar spiritual movement that went together with the beginning of Christianity, namely, the Gnostics; and that was the first thing actually that I saw. They were concerned with the problem of archetypes, and made a peculiar philosophy of it. Everybody makes a peculiar philosophy of it when he comes across it naively and doesn't know that those are structural elements of the unconscious psyche. The Gnostics lived in the first, second, and third centuries; and I wanted to know what was in between that time and today, when we suddenly are confronted by the problems of the collective unconscious which were the same 2,000 years ago, though we are not prepared to admit that problem. I was always looking for something in between, you know, something that would link that remote past with the present moment.

I found to my amazement that it is alchemy, that which is understood to be a history of chemistry. It is, one could almost say, anything *but* that. It was a peculiar spiritual movement or a philosophical movement. They called themselves philosophers, like Gnosticism.

And then I read the whole accessible literature, Latin and Greek. I studied it because it was enormously interesting. It is the mental work of 1,700 years in which there is stored up all they could make out about the nature of the archetypes. In a peculiar way that's true, it is not simple. Most of the texts are no more published since the Middle Ages, the last editions dated in the middle of the end of the sixteenth century, practically all in Latin; some texts are in Greek, not a few very important ones. That has given me no end of work, but the result was most satisfactory because it showed me the development of our unconscious relations to the collective unconscious and the variations our consciousness has undergone; why the being's unconscious is concerned with these mythological images.

For instance, such phenomena as in Hitler, you know. That is a psychical phenomenon, and we've got to understand these things. To me, of course, it was an enormous problem because it is a factor that has determined the fate of millions of European people, and of Americans. Nobody can deny that he has been influenced by the war. That was all Hitler's doing—and that's all psychology, our foolish psychology. But you only come to an understanding of these things when you understand the background from which it springs. It is just as though a terrific epidemic of typhoid fever were breaking out and you say, "That is typhoid fever—isn't that a marvelous disease!" It can take on enormous dimensions and nobody knows anything about it. Nobody takes care of the water supply, nobody thinks of examining the meat or anything like that; but everyone simply states, "This is a phenomenon." Yes, but one doesn't understand it.

Of course, I cannot tell you in detail about alchemy. It is the basis of our modern way of conceiving things, and therefore it is as if it were right under the threshold of consciousness. This is a wonderful picture of how the development of archetypes —the movement of archetypes—looks when you look upon them with broader perspective. Maybe from today you look back into the past and you see how the present moment has evolved out of the past. Alchemical philosophy—that sounds very curious, we should give it an entirely different name. Actually, it has a different name; it is also called Hermetic Philosophy, though, of course, that conveys just as little as the term alchemy. It was the parallel development, as Gnosticism was, to the conscious development of Christianity, of our Christian philosophy, of the whole psychology of the Middle Ages.

So you see, in our days we have such and such a view of the world, a particular philosophy, but in the unconscious we have a different one. That we can see through the example of alchemical philosophy that behaves to the medieval consciousness exactly like the unconscious behaves to ourselves. And we can reconstruct or even predict the unconscious of our day when we know what it has been yesterday.

Or, for instance, to take a more concise archetype like the archetype of the ford—the ford through a river. Now that is a whole situation. You have to cross a ford, you are in the water. There is an ambush or there is a water animal, say a crocodile or something like that. There is danger and something is going to happen. The question is how you escape. Now this is a whole situation that makes an archetype. And that archetype has now a suggestive effect upon you. For instance, you get into a situation; you don't know what the situation is; you suddenly are seized by an emotion or by a spell; and you behave in a certain way you have not foreseen at all—you do something quite strange to yourself.

DR. EVANS: Could this also be described as spontaneous?

DR. JUNG: Quite spontaneous. And that is done through that archetype that is concerned. Of course, we have a famous case in our Swiss history of the King Albrecht, who was murdered in the ford of the Reuss not very far from Zurich. His murderers were hiding behind him for the whole stretch from Zurich to the Reuss—quite a long stretch—and after deliberating, still couldn't come together about whether they wanted to kill the king or not. The moment the king rode into the ford, Johannes Parricida, the father-murderer, shouted, "Why do we let him abuse us?" And they killed him, because this was the moment they were seized; this was the right moment. So you see, when you have lived in primitive circumstances, in the primeval forest among primitive populations, then you know that phenomenon. You are seized with a certain spell and you do a thing that is unexpected.

Several times when I was in Africa, I went into such situations where I was amazed afterwards. One day I was in the Sudan and it was really a very dangerous situation, which I didn't recognize at the moment at all, but I was seized with a spell. I did something which I wouldn't have expected and I couldn't have intended.

You see, the archetype is a force. It has an autonomy, and it can suddenly seize you. It is like a seizure. So, for instance, falling in love at first sight, that is such a case. You have a certain image in yourself, without knowing it, of the woman—of *the* woman. You see that girl, or at least a good imitation of your type, and instantly you get the seizure and you are gone. And afterward you may discover that it was a hell of a mistake. You see, a man is quite capable, or is intelligent enough to see that the woman of his choice was no choice; he has been captured! He sees that she is no good at all, that she is a hell of a business, and he tells me so. He says, "For God's sake, doctor, help me to get rid of that woman." He can't though, and he is like clay in her fingers. That is the archetype. It has all happened because of the archetype of the anima, though he thinks it is all his soul, you know. It is like the girl—any girl. When a man sings very high, for instance, sings a high *C*, she thinks he must have a very wonderful spiritual character, and she is badly disappointed when she marries that particular "letter." Well, that's the archetype of the animus.

DR. EVANS: Now Dr. Jung, to be even a bit more specific, you have suggested that in our society—in all societies—there are symbols that in a sense direct or determine what a man does. Then you also suggest that somehow these symbols become inborn and, in part, inbred.

DR. JUNG: They don't become; they *are*. They are to begin with. You see, we are born into a pattern; we are a pattern. We are a structure that is preestablished through the genes.

DR. EVANS: To recapitulate then, the archetype is just a higher order of an instinctual pattern, such as your earlier example of a bird building a nest. Is that how you intended to describe it?

DR. JUNG: It is a biological order of our mental functioning, as, for instance, our biological-physiological function follows a pattern. The behavior of any bird or insect follows a pattern, and that is the same with us. Man has a certain pattern that makes him specifically human, and no man is born without it. We are only deeply unconscious of these facts because we live by all our senses and outside of ourselves. If a man could look into himself, he could discover it. When a man discovers it in our days, he thinks he is crazy and he may be crazy.

DR. EVANS: Now would you say the number of such archetypes are limited or predetermined, or can the number be increased?

DR. JUNG: Well, I don't know what I do know about it; it is so blurred. You see, we have no means of comparison. We know, you see, that there is a behavior, say like incest; or there is a behavior of violence, a certain kind of violence; or there is a behavior of panic, of power, etc. Those are areas, as it were, in which there are many variations. It can be expressed in this way or that way, you know. And they overlap, and often you cannot say where the one form begins or ends.

It is nothing concise, because the archetype in itself is completely unconscious and you only can see the effects of it. You can see, for instance, when you know a person possessed by an archetype; then you can divine and even prognosticate possible developments. This is true because when you see that the man is caught by a certain type of woman in a certain very specific way, you know that he is caught by the anima. Then the whole thing will have such and such complications and such and such developments because it is typical. The way the anima is described is exceedingly typical. I don't know if you know Rider Haggard's *She*, or *L'Atlantide* by Benoit. Those are anima types, and they are quite unmistakable. *C'est la femme fatale.*

DR. EVANS: To be more specific, Dr. Jung, you have used the concepts, "anima" and "animus," which you are now identifying in terms of sex—male or female. I wonder if you could elaborate perhaps even more specifically on these terms? Take the term "anima" first. Is this again part of the inherited nature of the individual?

DR. JUNG: Well, this is a bit complicated, you know. The anima is an archetypal form, expressing the fact that a man has a minority of feminine or female genes. That is something that doesn't appear or disappear in him, that is constantly present, and works as a female in a man.

As early as the sixteenth century, the Humanists had discoverd that man had an anima; each man carried his female within himself, they said. It is not a modern invention. The same is the case with the animus. It is a masculine image in a woman's mind which is sometimes quite conscious, sometimes not quite conscious; but it is called into life the moment that woman meets a man who says the right things. Then because he says it, it is all true and he is *the* fellow, no matter what he is. Those are particularly well-founded archetypes, those two. And there you can lay hands on the basis, as it were, of the archetype. They are extremely well-defined.

Dreams
and the
Personal
Unconscious

DR. EVANS: Dr. Jung, to pursue our discussion of the unconscious further, let us take the particular situation of a dream and its interpretation. It is my understanding that, in your view of the unconscious, what you would find in the dream would not necessarily be an image or symbol of what has happened in the past to the individual.

DR. JUNG: Oh no! It just is a symbol of the—*symbol*, you see, that is a special term—it is the manifestation of the situation of the unconscious, looked at from the unconscious. You see, I tell you, for instance, something which is my personal subjective view. And if I ask myself, "Now are you really quite convinced of it?", well, I must admit I have certain doubts. There are certain doubts, not in the moment when I tell you, but these doubts are in the unconscious, and when I have a dream about it, these doubts come to the forefront. That is the way the unconscious looks at the thing. It is just as though the unconscious says, "It is all very well what you are stating, *but* you omit entirely such and such a point."

DR. EVANS: Now if the unconscious acts on the present situation, looking at this in broad motivational terms, this effect of the unconscious is not something which is a result of repression in the way the orthodox psychoanalyst looks at it at all. Then . . .

DR. JUNG: It may be, you know, that what the unconscious has to say is so disagreeable that one prefers not to listen, and in most cases people would be probably less neurotic if they could admit the things. But these things are always a bit difficult or disagreeable, inconvenient, or something of the sort; so there is always a certain amount of repression, but that is is not the main thing.

The main thing is that they are really unconscious. If you are unconscious about certain things that ought to be conscious, then you are dissociated. Then you are a man whose left hand never knows what the right is doing, and counteracts or interferes with the right hand. Now such a man is hampered all over the place.

In 1916 I wrote a disquisition on the relation between the ego and the unconscious. There I tried to formulate the experiences that are regularly observable in cases where consciousness is exposed to unconscious data, to interferences or intrusions; where the unconscious is considered as an autonomous factor that has to be taken seriously; where one doesn't say anymore or undervalue the unconscious by assuming that it is nothing but a discarded remnant of consciousness. It is a factor in its own dignity, and a very important factor, because it can create the most horrible disturbances.

When I wrote that pamphlet in 1916, it was published in French and nobody understood it. I saw, however, the reason why nobody understood it. It was because nobody had had a similar experience, because the question hadn't been pursued to such an end. To pursue that end, one has to take the unconscious seriously and consider it as a real factor that can determine human behavior to a very considerable degree.

DR. EVANS: Looking at the unconscious in this way, as you say, "If it's unconscious, how do we know about it?" But just as an illustration let us consider a particular individual, for example, one who has been brought up in a culture such as India. Would this individual, if we could examine his unconscious, be in many respects similar to the unconscious of a particular individual who, let us say, has lived in Switzerland all his life? You spoke earlier about these universals. Would there be quite a lot of equivalence between the unconscious of a particular individual who was raised in one culture, and another individual who was raised in an entirely different culture?

DR. JUNG: Well, that question is also complicated because when we speak of the unconscious, we almost should say, "Which unconscious?" Namely, is it that personal unconsciousness which is characteristic for a certain person, for a certain individual?

DR. EVANS: You have talked in your writings about a personal unconsciousness as being one kind of unconscious.

DR. JUNG: Yes. In treatment, for instance, the treatment of neurosis, you have to do with that personal unconsciousness for quite a while, and then only dreams come that show that the collective unconscious is touched upon. As long as there is material of a personal nature, you have to deal with the personal unconscious. But when you get to a question, to a problem which is no more merely personal but also collective, you get collective dreams.

DR. EVANS: The distinction between the personal unconscious and the collective unconscious, then, is that the personal would be more involved with the immediate life of the individual, and the collective would be universal—the unconscious realm composed of elements which are the same in all men?

DR. JUNG: It would be collective. For instance, in all of us, the psyche has collective problems, collective convictions, etc. We are very much influenced by them, and there are examples to prove it. You belong to a certain political party, or to a certain confession; that can be a serious determinant of your behavior. Now if there arises a matter of personal conflict, the collective unconscious isn't touched upon. There is no question and it doesn't appear. But the moment you transcend your personal sphere and come to your own personal "determinants"—say to a political question, or to any other social question which really matters to you—then you are confronted with a collective problem; then you have collective dreams.

DR. EVANS: Another very interesting concept or idea in your work is the "persona." This seems to be highly relevant to the daily living of the individual. I wonder if you would mind elaborating a bit more about how you construe this term, "persona."

DR. JUNG: It is a practical concept we need in elucidating people's relations. I noticed with my patients, particularly with people that are in public life, that they have a certain way of presenting themselves. For instance, take a doctor. He has a certain way; he has good bedside manners, and he behaves as one expects a doctor to behave. He may even identify himself with it, and believe that he is what he appears to be. He must appear in a certain form, or else people won't believe that he is a doctor. And so when one is a professor, he is also supposed to behave in a certain way so that it is plausible that he is a professor. So the persona is partially the result of the demands society has.

On the other side, it is a compromise with what one likes to be, or as one likes to appear. Take, for instance, a parson. He also has his particular manner and, of course, runs into the general societal expectations; but he behaves also in another way, combined with his persona that is forced upon him by society in such a way that his fiction of himself, his idea about himself, is more or less portrayed or represented.

So the persona is a certain complicated system of behavior which is partially dictated by society and

partially dictated by the expectations or the wishes
one nurses oneself. Now this is not the real personality. In spite of the fact that people will assure
you that that is all quite real and quite honest, yet it
is not. Such a performance of the persona is quite all
right, as long as you know that you are not identical to
the way in which you appear; but if you are unconscious
of this fact, then you get into sometimes very disagreeable conflicts. Namely, people can't help noticing
that at home you are quite different from what you
appear to be in public. People who don't know it
stumble over it in the end. They deny that they are
like that, but they *are* like that; they are it. Then you
don't know—now which is the real man? Is he the man
as he is at home or in intimate relations, or is he the
man that appears in public?

It is a question of Jekyll and Hyde. Occasionally
there is such a difference that we would almost be
able to speak of a double personality, and the more
that it is pronounced, the more people are neurotic.
They get neurotic because they have two different
ways; they contradict themselves all the time, and
inasmuch as they are unconscious of themselves, they
don't know it. They think they're all one, but everybody sees that they are two. Some know him only
from one side; others know him only from the other
side. And then there are situations that clash, because
the way you are creates certain situations with people
in your relations, and these two situations don't chime;
in fact, they are just dishonest. And the more that is
the case, the more the people are neurotic.

DR. EVANS: Actually, would you say that the individual may even have more than two personas?

DR. JUNG: We can't afford very well to play more than two roles, but there are cases where people have up to five different personalities. In cases of dissociation of personality, for instance, the one person— call him A—doesn't know of the existence of the person B, but B knows of A. There may be a third personality, C, that doesn't know of the two others. There are such cases in the literature, but they are rare.

DR. EVANS: Very rare.

DR. JUNG: In ordinary cases, it's just an ordinary dissociation of a personality. One calls that a systematic dissociation, in contradistinction to the chaotic or unsystematic dissociation you find in schizophrenia.

DR. EVANS: What is the difference between the term "ego" as you see it and the term "persona?"

DR. JUNG: Well, you see, the ego is supposed to be the representative of the real person. For instance, in the case where B knows of A, but A doesn't know of B, one would say the ego is more on the side of B, because the ego has a more complete knowledge, and A is a split-off personality.

DR. EVANS: You also use the term "self." Now the word "self"—does it have a different meaning than "ego" or "persona"?

DR. JUNG: Yes. When I say "self," then you mustn't think of "I, myself," because that is only your empirical self, and that is covered by the term "ego"; but when it is a matter of self, then it is a matter of a personality that is more complete than the ego, because the ego only consists of what you are conscious of, what you know to be yourself. For instance, let's take our example again of B who knows A, but A who doesn't know B. B is relatively in the position of the self; namely, the self is, on the one side the ego, on the other side, the unconscious personality (which is in the possession of everybody) not in possession. Very often it is just the other way around; the unconscious is in the possession of consciousness. That is a different case. Now you see, while I am talking, I am conscious of what I say; I am conscious of myself, yet only to a certain extent. Quite a lot of things happen. For instance, I make gestures—I'm not conscious of them. They happen unconsciously. You can see them. I may say or use words and can't remember at all having used those words, or even at the moment I am not conscious of them. So any amount of unconscious things occur in my conscious condition. I'm never wholly conscious of myself. While I am trying, for instance, to elaborate an argument, at the same time there are unconscious processes that continue, perhaps a dream which I had last night; or a part of myself thinks of God knows what, of a trip I'm going to take or of such and such people I have seen. Or say when I am writing a paper, I am continuing writing that paper in my mind without knowing it. You can discover these things, say in dreams; or if you are clever, in the immediate observation of the individual. Then you see in the gestures or in the expression in the face that there is what one calls *une arrière pensée,* something behind consciousness. You have finally the feeling, well, that that man has something up his sleeve, and you can even ask him, "What are you really thinking of? You are thinking all the time something else." Yet he is not conscious of it. Or he may be.

There are, of course, great individual differences. There are individuals who have an amazing knowledge of themselves, of the things that go on in themselves. But even those people wouldn't be capable of knowing what is going on in their unconscious. For instance, they are not conscious of the fact that while they live a conscious life, all the time a myth is played in the unconscious, a myth that extends over centuries, namely, archetypal ideas—a stream of archetypal ideas that goes on through an individual through the centuries. Really it is like a continous stream, one that comes into the daylight in the great movements, say in political movements or in spiritual movements. For instance, in the time before the Reformation, people dreamt of the great change. That is the reason why such great transformations could be predicted.

If somebody is clever enough to see what is going on in people's minds, in the unconscious mind, he will be able to predict. For instance, I could have predicted the Nazi rising in Germany through the observation of my German patients. They had dreams in which the whole thing was anticipated, and with considerable detail. And I was absolutely certain—in the years before Hitler, before Hitler came in the beginning; I could say the year, in the year 1919—I was sure that something was threatening in Germany, something very big, very catastrophic. I only knew it through the observation of the unconscious.

There is something very particular in the different nations. It is a peculiar fact that the archetype of the anima plays a very great role in Western literature, French and Anglo-Saxon. Not in Germany; there are exceedingly few examples in German literature where the anima plays a role.

You know, that simply comes of the fact that not one woman is buried unless she is buried as *alt Kaminfegersgattin* ["Old chimney sweep's wife"], at least. She must have a title; otherwise she hasn't existed. And so it is just as if—now mind you, this is a bit drastic, but it illustrates my point—in Germany there really are no women. There is Frau Doktor, Frau Professor, the grandmother, the mother-in-law, the daughter, the sister. *La femme n'existe pas*—no woman. That is the idea, you see. Now that is an enormously important fact which shows that in the German mind there is going on a particular myth, something very particular and psychologists really should look out for these things. They prefer to think that I am a prophet. Ha!

DR. EVANS: This is of course a very interesting and remarkable set of statements here. How would you look at Hitler in this light? Would you see him as a personification, as a symbol of *the father?*

DR. JUNG: No, not at all. I couldn't possibly explain that very complicated fact that Hitler represents. It is too complicated. You know, he was a hero figure, and a hero figure is far more important than any fathers that have ever existed. He was a hero in the German myth and mind you, a religious hero. He was a savior; he was meant to be a savior. That is why they put his photo upon the altars even. And that's why somebody declared on his tombstone that he was happy that his eyes had beheld Hitler, and that now he could lie in peace. Hitler was just a hero myth.

DR. EVANS: To get back more specifically to the idea of the self . . .

DR. JUNG: The self is merely a term that designates the whole personality. The whole personality of man is indescribable. His consciousness can be described; his unconscious cannot be described, because the unconscious—and I repeat myself—is always unconscious. It is really unconscious; he really does not know it. And so we don't know our unconscious personality. We have hints and certain ideas, but we don't know it really.

Nobody can say where man ends. That is the beauty of it, you know. It is very interesting. The unconscious of man can reach . . . God knows where. There we are going to make discoveries.

DR. EVANS: What seems to be a very fundamental part of your writing and one of your main ideas is reflected in the term "mandala." How does this fit into the context of our discussion of the self?

DR. JUNG: Mandala is just one typical archetypal form. It is what they called in alchemy, the *quadratura circuli,* the square in the circle, or the circle in the square. It is an age-old symbol that goes right back to the prehistory of man. It is all over the earth and it either expresses the Deity or the self; and these two terms are psychologically very much related, which doesn't mean that I believe that God is the self or that the self is God. I made the statement that there is a psychological relation, and there is plenty of evidence for that.

It is a very important archetype. It is the archetype of inner order; and it is always used in that sense, either to make arrangements of the many, many aspects of the universe—a world scheme—or to arrange the complicated aspects of our psyche into a scheme. It expresses the fact that there is a center and a periphery, and it tries to embrace the whole. It is the symbol of wholeness. So you see, in a moment during a patient's treatment when there is a great disorder and chaos in a man's mind, then the symbol can appear, as in the form of a mandala in a dream, or when he makes imaginary fantastical drawings, or something of the sort.

A mandala spontaneously appears as a compensatory archetype during times of disorder. It appears, bringing order, showing the possibility of order and centeredness. It means a center which is not coincident with the ego, but with the wholeness—it is wholeness— the wholeness which I call "the self"; this is the term for wholeness. I am not whole in my ego as my ego is but a fragment of my personality. So you see, the center of a mandala is not the ego.

It is the whole personality, the center of the whole personality, and the very great role that it plays can be seen, for instance, in the culture of the East, past and present. In the Middle Ages it played an equally great role for the West, but there it has been lost now and is thought of as a mere sort of allegorical, decorative motif. As a matter of fact, however, it is highly important and highly autonomous, a symbol that appears in dreams, etc., and in folklore. We could say that it is the main archetype.

III. MAJOR THEORIES AND CONCEPTS

Perhaps Jung's most widely known contribution is his *type* theory, in which he sets up the dichotomy of the introvert and the extrovert. Jung was quite distressed over the misinterpretation of his ideas by Americans, and was quite aware that his introvert-extrovert typology had been the recipient of much of this misinterpretation. In these interviews, Jung reflects his lack of patience with this distortion of his intended meaning and usage of these terms.

He explains in great detail the interrelationship that exists between what he refers to as the four functions—thinking, feeling, sensation, and intuition—and what he designates as the introversion and extroversion orientations. Particularly difficult to understand fully is his introvert-intuitive type, so he gives some intriguing case material to illustrate this orientation in an individual.

In discussing his motivational conceptions, primarily his view of the libido, he explains his unique concept of energy as manifested in the individual. He also seems to support the importance of historical factors in understanding the individual, but not to the exclusion of emphasis on understanding the current events influencing the person, viz., the importance of a field approach.

Introverts and Extroverts

DR. EVANS: Dr. Jung, another set of ideas, original with you and very well-known to the world, center around the terms "introversion" and "extroversion." I know that you are aware that these terms have now become so widely known that the man on the street is using them constantly in describing members of his family, his friends, and so on. They have become probably the psychological concepts most often used by the layman today.

DR. JUNG: Like the word "complex"—I invented it too, you know, from the association experiments—this is simply practical, because there are certain peo-

ple who definitely are more influenced by their surroundings than by their own intentions, while other
people are more influenced by the subjective factor.
Now you see, the subjective factor, which is very
characteristic, was understood by Freud as a sort of
pathological autoerotism. This is a mistake. The psyche
has two conditions, two important conditions. The
one is environmental influence and the other is the
given fact of the psyche as it is born.

The psyche is by no means *tabula rasa*, but a
definite mixture and combination of genes, which
are there from the very first moment of our life; and
they give a definite character, even to the little child.
That is a subjective factor, looked at from the outside.
Now if you look at it from the inside, then it is just
so as if you would observe the world. When you observe the world, you see people; you see houses; you
see the sky; you see tangible objects. But when you
observe yourself within, you see moving images, a
world of images generally known as fantasies.

Yet these fantasies are facts. You see, it is a fact
that the man has such and such a fantasy; and it is
such a tangible fact, for instance, that when a man
has a certain fantasy, another man may lose his life,
or a bridge is built. These houses were all fantasies.
Everything you do here—all this, everything—was fantasy to begin with, and fantasy has a proper reality.
That is not to be forgotten; fantasy is not nothing. It
is, of course, not a tangible object; but it is a fact
nevertheless.

Fantasy is, you see, a form of energy, despite the
fact that we can't measure it. It is a manifestation of

something, and that is a reality. That is a reality, like for instance, the Peace Treaty of Versailles, or something like that. It is no more; you can't show it, but it has been a fact. And so psychical events are realities. And when you observe the stream of images within, you observe an aspect of the world, of the world within. Because the psyche, if you understand it as a phenomenon that takes place in so-called living bodies, is a quality of matter, as our bodies consist of matter. We discover that this matter has another aspect, namely, a psychic aspect. And so it is simply the world from within, seen from within. It is just as though you were seeing into another aspect of matter. That is an idea that is not my invention. The old credos already talked of the *spiritus insertus atomis*—namely, the spirit that is inserted in atoms. That means psychic is a quality that appears in matter. It doesn't matter whether we understand it or not, but this is the conclusion we come to if we draw conclusions without prejudices.

And so you see, the man who is going by the external world, by the influence of the external world—say society or sense perceptions—thinks that he is more valid, you know, because this is valid, this is real; and the man who goes by the subjective factor is not valid, because the subjective factor is nothing. No, that man is just as well based, because he bases himself upon the world from within. And so he is quite right even if he says, "Oh, it is nothing but my fantasy." And of course, that is the introvert, and the introvert is always afraid of the external world. This he will tell you when you ask him. He will be apologetic about it, he will say, "Yes, I know, those are my fantasies." And he has always resentment against the world in general.

Particularly America is extroverted like hell. The introvert has no place, because he doesn't know that he beholds the world from within. That gives him dignity, and that gives him certainty, because nowadays particularly, the world hangs by a thin thread, and that thread is the psyche of man. Assume that certain fellows in Moscow lose their nerve or their common sense for a bit; and the whole world is in fire and flames. Nowadays we are not threatened by elemental catastrophies. There is no such thing in nature as an H-bomb; that is all man's doing. *We* are the great danger. The psyche is the great danger. What if something goes wrong with the psyche? And so it is demonstrated to us in our days what the power of psyche is, how important it is to know something about it. But we know nothing about it. Nobody would give credit to the idea that the psychical processes of the ordinary man have any importance whatever. One thinks, "Oh, he has just what he has in his head; he is all from his surroundings; he is taught such and such a thing, believes such and such a thing, and particularly if he is well housed and well fed, then he has no ideas at all." And that's the great mistake, because man is just that or which he is born, and he is not born as *tabula rasa* but as a reality.

So I began an examination of the human attitudes, namely, how our consciousness functions. I couldn't help seeing, for instance, the difference between Freud and Adler, a typical difference. The one assumed that things evolve along the line of the sex instinct. The other assumed that things evolve along the line of power drive. And there I was, in between the two. I could see the justification of Freud's view, and also could see the same for Adler; and I knew that there were plenty of other ways in which things could be envisaged. And so I considered it my scientific duty to examine first the condition of the human consciousness, that which is the originator of ways of envisaging. It is the factor that produces attitudes, conscious attitudes, towards certain phenomena. So when you know, for instance, that there are people who see the difference between red and green, can you take it for a fact that everybody sees that difference? Not at all. There are cases of color blindness. You know, the one sees this; the other sees that.

Thus, I tried to find out what the principal differences were. That is the book about the types. I saw first the introverted and extroverted attitudes, then certain functional aspects, and then which of the four functions is predominant.

DR. EVANS: Of course, one of the very common misconceptions, at least in my opinion, of your work among some of the writers in America is that they have characterized your discussion of introversion and extroversion as suggesting that the world is made up of only two kinds of people, introverts and extroverts. I'm sure you have been aware of this. Would you like to comment on it? In other words, do you perceive of the world as one made up only of people who are extreme introverts and people who are extreme extroverts?

DR. JUNG: Bismarck once said, "God preserve me from my friends; with my enemies I can deal alone." You know how people are. They have a catchword, and then everything is schematized along that word. There is no such thing as a pure extrovert or a pure introvert. Such a man would be in the lunatic asylum.

Those are only terms to designate a certain penchant, a certain tendency. For instance, the tendency to be more influenced by environmental influences, or more influenced by the subjective fact—that's all. There are people who are fairly well-balanced who are just as much influenced from within as from without, or just as little. And so with all the definite classifications, you know, they are only a sort of point to refer to, points for orientation. There is no such thing as a schematic classification.

Often you have great trouble even to make out to what type a man belongs, either because he is very well-balanced or he is very neurotic. The last one is hard because when you are neurotic, then you have always a certain dissociation of personality. And then too, the people themselves don't know when they react consciously or when they react unconsciously. So you can talk to somebody, and you think he is conscious. He knows what he says, and to your amazement you discover after a while that he is quite unconscious of it, doesn't know it.

It is a long and painstaking procedure to find out of what a man is conscious and of what he is not conscious, because the unconscious plays in him all the time. Certain things are conscious, certain things are unconscious; but you can't always tell. You have to ask people, "Now are you conscious of what you say?" Or, "Did you notice?" And you discover suddenly that there are quite a number of things that he didn't know at all. For instance, certain people have many reasons; everybody can see them. They themselves don't know it at all.

DR. EVANS: Then this whole matter of extremes —introvert and extrovert—you say is a schematic approach, a frame of reference?

DR. JUNG: My whole scheme of typology is merely a sort of orientation. There is such a factor as introversion; there is such a factor as extroversion. The classification of individuals means nothing at all. It is only the instrumentality, for what I call "practical psychology," used to explain, for instance, the husband to a wife, or vice versa.

It is very often the case, for instance—I might say it is almost a rule, but I don't want to make too many rules in order not to be schematic—that an introvert marries an extrovert for compensation, or another type marries a countertype to complement himself. For example, a man who has made a certain amount of money is a good businessman, but he has no education. His dream is, of course, a grand piano at home and being around artists, painters, or singers or God knows what, and intellectual people; and he marries accordingly a wife of that type, in order to have that too. She has it, and she marries him because he has a lot of money. These compensations go on all the time. When you study marriages, you can see it easily. And, of course, we analysts have to deal a lot with marriages, particularly those that go wrong, because the types are too different sometimes and they don't understand each other at all.

You see, the main values of the extrovert are anathema to the introvert, and he says, "To hell with the world, I think." His wife interprets this as his megalomania. But it is just as if an extrovert said to an introvert, "Now, look here fellow; these here are the facts; this is reality." And he's right! And the other says, "But *I* think, *I* hold—," and that sounds like nonsense to the extrovert because he doesn't know that the other one, without knowing it, is beholding an inner world, an inner reality; and that other one may be right, as he may be wrong, even if he found himself upon God knows what solid facts. Take, for instance, the interpretation of statistics. You can prove almost anything with statistics. What is more a fact than a statistic?

Thinking, Feeling, Sensation, and Intuition Types

DR. EVANS: Of course, tied in with your typology of introversion-extroversion, we know of your four functions of thinking, feeling, sensation, and intuition. It would be very interesting to hear some expansion of the meaning of these particular terms as related to the introvert-extrovert orientations.

DR. JUNG: Well, there is a quite simple explanation of these terms, and it shows at the same time how I arrived at such a typology. Namely, sensation tells you that there is something. Thinking, roughly speaking, tells you what it is. Feeling tells you whether it is agreeable or not, to be accepted or not, accepted or rejected. And intuition—there is a difficulty because you don't know ordinarily how intuition works. When a man has a hunch, you can't tell exactly how he got that hunch, or where that hunch came from. It is something funny about intuition.

I will tell you a little story. I had two patients. The man was a sensation type, and the woman was an intuitive type. Of course, they felt attraction, so they took a little boat and went out to the lake of Zurich. And there at the lake were those birds that dive after fish, you know, that come up again after a certain time, only you can't tell where they will come up. My two patients began to bet about who would be the first to see the bird. Now you would think that the one who observes reality very carefully, the sensation type, would win out. Not at all. The woman won the bet completely. She was beating him on all points because by intuition she knew it before. How is that possible? You know, you can really find out how it works by finding the intermediate links. It is a perception by intermediate links, and you only get the result of that whole chain of associations. Sometimes you succeed in finding out, but more often you don't.

My definition then is that intuition is a perception, by ways or means of the unconscious. That is as near as I can get it. This is a very important function, because when you live under primitive conditions, a lot of unpredictable things are likely to happen. There you need your intuition because you cannot possibly tell by your sense perceptions what is going to happen. For instance, you are walking in primeval forests. You can only see a few steps ahead, and perhaps you go by the compass. You don't know what there is ahead; it is uncharted country. If you use your intuition, then you have hunches; and when you live under such primitive conditions, you instantly are aware of hunches. There are places that are favorable and there are places that are not favorable. You can't tell for your life what it is, but you better follow these hunches, because anything can happen, quite unforeseen things. For example, at the end of a long day you approach a river. You had not known that there was a river there, but unexpectedly you come upon this river. For miles there is no human habitation. You cannot swim across; it is full of crocodiles. So what? Such an obstacle hasn't been foreseen. It may be though that you have a hunch that you should remain in the least likely spot and wait for the following day; or that you should build a raft or something of the sort; or just a hunch that you should wait and look out for possibilities.

You can also have intuition—this constantly happens—in our jungle called a city. You can have a hunch that something is going wrong, particularly when you are driving an automobile. For instance, it is a day when nurses appear in the street. At one corner a nurse runs in front of the automobile. Now they always try to get something interesting, like a suicide, you know; to be run over, that's more marvelous apparently. And then you get a peculiar feeling really, for at the next corner a second nurse runs in front of the automobile. A multiplicity of cases, that is the rule, that such chance happenings come in groups.

So you see, we have constantly warnings or hints, that consist partly in a slight feeling of uneasiness, uncertainty, fear. Now under primitive circumstances you would pay attention to these things; they would mean something. With us in our man-made, absolutely, apparently, safe conditions we don't need that function so very much; yet we have seen it and used it. So you will find that the intuitive types, for instance, amongst bankers, Wall Street men, they follow their hunches, and so do gamblers of all descriptions. You find the type very frequently among doctors because it helps them in their prognoses. Sometimes a case can look quite normal, as it were, and you don't foresee any complications; yet an inner voice says, "Now you look out here, because there is something not quite all right."

You cannot tell why or how, but we have a lot of subliminal perceptions, sense perceptions, and from these we probably draw a great many of our intuitions. But that is perception by the way of the unconscious, and you can observe that with intuitive types. You see, intuitive types very often do not perceive by their eyes or by their ears; they perceive by intuition. For instance, once it happened that I had a woman patient in the morning at nine o'clock. Now I often smoke my pipe and have a certain smell of tobacco in the room, or a cigar. When she arrived, she said, "But you begin earlier than nine o'clock; you must have somebody at eight o'clock." And I said, "How do you know?" You see, there had been a man there at eight o'clock already. And she said, "Oh, I just had a hunch that there must have been a gentleman with you this morning." I said, "How do you know it was a gentleman?" And she said, "Oh well, I just had the impression; the atmosphere was just like a gentleman here." All the time, you know, the ashtray was under her nose, and there was a half smoked cigar but she wouldn't notice it. So you see, the intuitive is a type that doesn't see the stumbling block before his feet, but he smells a rat for ten miles.

DR. EVANS: How did you develop your conceptualizations of these four functions?

DR. JUNG: Now mind you, these four functions were not a scheme I had simply invented and applied to psychology. On the contrary, it took me quite a long time to discover them. Take the thinking type for example, as I thought my type to be. Of course, that is human, is it not? There are other people who decide the same problems that I am faced with and have to decide about, but they make their decisions in an entirely different way. They look at things in an entirely different way; they have entirely different values. They are, for instance, feeling types.

And so, after a while I discovered that there are intuitive types. They gave me much trouble. It took me over a year to become somewhat clear about the existence of the intuitive types. And the last, and the most unexpected, was the sensation type. And only later I saw that those are naturally the four aspects of conscious orientation.

You see, you get your orientation, you get your bearings, in the chaotic abundance of impressions through the four functions, these four aspects of total human orientation. If you can tell me any other aspect

by which you get your orientation, I'm very grateful.
I haven't found more and I tried. But those are the
four that covered the thing.

For instance, the intuitive type, to discuss it once
again, which is very little understood, has a very important function because he is the one going by
hunches. He sees around corners; he smells a rat a
mile away. He can give you perception and orientation in a situation where your senses, your intellect,
and your feelings are no good at all. When you are in
an absolute fix, an intuition can show you the hole
through which you can escape. This is a very important function under primitive conditions or wherever you are confronted with vital issues that you cannot master by rules or logic.

So, through the study of all sorts of human types,
I came to the conclusion that there must be many
different ways of viewing the world through these
type orientations—at least sixteen, and you can just as
well say 360. You can increase the number of guiding
or underlying principles, but I found that the most
simple way is the way I told you, the division by four,
the simple and natural division of the circle. I didn't
know the symbolism of this particular classification.
Only when I studied the archetypes did I become
aware that this is a very important archetypal pattern that plays an enormous role.

DR. EVANS: Do you make a distinction between
an intuitive extrovert and an intuitive introvert?

DR. JUNG: Yes, all those types cannot be alike.

DR. EVANS: More specifically, what would be an

example of the difference between an intuitive extrovert and an intuitive introvert?

DR. JUNG: Well, you have chosen a somewhat difficult case, because one of the most difficult types is
the intuitive introvert. The intuitive extrovert you find
in all kinds of bankers, gamblers, etc., which is quite
understandable. But the introvert variety is more
difficult because he has intuitions as to the subjective
factor, namely the inner world; and, of course, that is
now very difficult to understand because what he sees
are most uncommon things, things which he doesn't
like to talk about if he is not a fool. If he did, he would
spoil his own game by telling what he sees, because
people won't understand it.

For instance, once I had a patient, a young
woman about twenty-seven or twenty-eight. Her first
words when I had seated her were, "You know, doctor,
I come to you because I've a snake in my abdomen."
What! "Yes, a black snake coiled up right in the
bottom of my abdomen." I must have made a rather
bewildered face at her, so she said, "You know that I
don't mean it literally." I then replied, however, "If
you say it was a snake, it was a snake."

In a later conversation with her, which took place
about in the middle of her treatment, treatment that
only lasted for ten consultations, she reminded me of
something she had foretold me. She had said, "I will
come ten times and then it will be all right," to which
I responded with the question, "How do you know?"
"Oh, I've got a hunch," she said. Now at about the
fifth or sixth hour she said, "Doctor, I must tell you

that the snake has risen; it is now about here." A hunch.

Then on the tenth day I said, "Now this is our last hour, and do you feel cured?" Just beaming, she replied, "You know, this morning it came up, came out of my mouth, and the head was golden." Those were her last words.

When it comes to reality now, that same girl came to me because she couldn't hear the step of her feet anymore, because she walked on air, literally. She couldn't hear it, and that frightened her. When I asked for her address, she said, "Oh, Pension so-and-so. Well, it is not just called a pension, but it is a sort of pension." I had never heard of it. "I have never heard of that place," I said. She replied, "It is a very nice place. Curiously enough, there are only young girls there; very nice and very lively young girls, and they have a merry time. I often wish they would invite me to their merry evenings." And I said, "Do they amuse themselves all alone?" "No," she replied, "there are plenty of young gentlemen coming in; they have a beautiful time, but they never invite me." It turned out that this was a private brothel. She was a perfectly decent girl from a very good family, not from here. She had found that place, I don't know how, and she was completely unaware that they were all prostitutes. I said, "For heaven's sake, you fell into a very tough place; you hasten to get out of it."

That was her sensation; she didn't see reality, but she had hunches like everything. Such a person cannot possibly speak of her experiences because everybody would think she was absolutely crazy. I myself was quite shocked, and I thought, "For heaven's sake, is that case a schizophrenia?" You don't normally hear that kind of speech; but she assumed that the old man, of course, knew everything and did understand such kind of language.

So you see, if the introverted intuitive would speak what he really perceives, practically no one would understand him; he would be misunderstood. Thus they learn to keep things to themselves. You hardly ever hear them talking of these things. In a way, that is a great disadvantage, but in another way it is an enormous advantage that these people do not speak of their experiences, both their inward experiences and those which occur in human relations. For instance, they may come into the presence of somebody they don't know, not from Adam, and suddenly they may have inner images. Now these inner images may give them a great deal of information about the psychology of that person they have just met. That is typical of cases that often happen. They suddenly know an important piece out of the biography of that person, and if they did not keep things to themselves, they would tell the story. Then the fat would be in the fire! So the intuitive introvert has in a way a very difficult life, although it is a most interesting one. It is often quite difficult to get into their confidence.

DR. EVANS: Yes, because they are afraid people will think . . .

DR. JUNG: They are sick. The things that they hint at are interesting to them, are vital to them, and are utterly strange to the ordinary individual. A psychologist, however, should know of such things.

When people make a psychology, as a psychologist ought to do, it is the very first question—is he introverted or is he extroverted? The psychologist must look at entirely different things. Is he the sensation type; is he the intuitive type; is he the thinking or the feeling type?

These things are complicated. They are still more complicated because the introverted thinker, for instance, is compensated by extroverted feeling, inferior, archaic, extroverted feeling. So an introverted thinker may be very crude in his feeling, like for instance the introverted philosopher who is always carefully avoiding women may be married by his cook in the end.

DR. EVANS: So we can take your introvert-extrovert orientations and describe a number of types; the sensation-introvert and extrovert types, the feeling-introvert and extrovert types, thinking-introvert and extrovert types, and the intuitive-introvert and extrovert types. In each case these combinations do not represent a concrete category but simply, as you have indicated, a model that can be helpful in understanding the individual.

DR. JUNG: It is just a sort of skeleton to which you have to add the flesh. One could say that it is like a country mapped out by triangulation points, which doesn't mean that the country consists of triangulation points; that is only in order to have an idea of the distances. And so it is a means to an end.

It only makes sense as a scheme when you deal with practical cases. For instance, if you have to explain an introverted-intuitive husband to an extrovert wife, it is a most painstaking affair because, you see, an extrovert-sensation type is furthest away from the inner experience and the rational functions. She adapts and behaves according to the facts as they are, and she is always caught by those facts. She, herself, is those facts.

But if the introvert is intuitive, to him that is hell, because as soon as he is in a definite situation, he tries to find a hole where he can get out. To him, every given situation is just the worst that can happen to him. He is pinched and feels he is caught, suffocated, chained. He must break those fetters, because he is the man who will discover a new field. He will plant that field, and as soon as the new plants are coming up, he's done; he's over and no more interested. Others will reap what he has sown. When those two marry [the extrovert-sensation and the introvert-intuitive], there is trouble, I can assure you.

Motivation and Psychic Individuation

DR. EVANS: One question which is quite important, as we attempt to understand the individual, centers around the problem of motivation, why the person does what he does. To some degree you have already talked about this when you discussed archetypes. However, to go further into this problem, earlier when we discussed the libido, that which Freud considered a psychic, sexual energy, you may recall your suggestion that it was more than just sexual energy. You suggested that it could be something much broader. You have certain principles concerning psychic energy which are very provocative, and one of these principles, I believe you refer to as the principle of "entropy."

DR. JUNG: Well, I only allude to it. The main point is to take the standpoint of energetics as applied to psychical phenomena. Now with psychical phenomena you have no possibility to measure exactly, so it always remains a sort of analogy.

Freud uses the term "libido" in the sense of sexual energy, and that is *not* quite correct. If it is sexual, then it is a power, like electricity or any other form of manifestation of energy. Now energy is a concept by which you try to express the analogies of all power manifestations; namely, that they have a certain quantity, a certain intensity, and there is a flow in one direction, viz., to the ultimate suspension of the opposites. Low-high-height—a lake on a mountain flows down until all the water is down, you know, then it is finished. And you see something similar in the case in psychology.

We get tired from intellectual work or from consciously living, and then we must sleep to restore our powers. Then by sleeping through the night, it is as if the water were pumped from a lower level to a higher level, and we can work again the next day. Of course, that simile is limping too, so it is only in an analogous way that we use the term "energy."

I used it because I wanted to express the fact that the power manifestation of sexuality is not the only power manifestation. You have a number of drives, say the drive to conquer or the drive to be aggressive, or any number of others. There are many forms. For instance, take animals, the way they build their nests, or the urge of the traveling birds that migrate. They all are driven by a sort of energy manifestation, and the meaning of the word "sexuality" would be entirely gone if all these different urges and drives were included in its definition. Freud himself says that this is not applicable everywhere, and later on he corrected himself by assuming that there are also ego drives. That is something else, another manifestation.

Now in order not to presume or to prejudice things, I speak simply of energy, a quantity of energy that can manifest itself via sexuality or via any other instinct. That is the main feature, not the existence of one single power, because that is not warrantable.

DR. EVANS: Many approaches to motivation in our academic psychology today emphasize what is sometimes referred to as a biocentric theory. It suggests that the individual is born with certain inborn physiological, self-preserving types of drives, such as the drive for hunger, thirst, etc. Sex is just one of them. In the case of all these drives, however, their satisfaction is necessary to the maintenance of the organism.

Then as the individual is influenced by reality and the culture in which he lives, these primary drives are modified in terms of the society in which he functions. For example, as a result of specific cultural influences, the general hunger drive is supplemented by a specific urge for certain kinds of food. Later, if this is important in the culture in which he lives, he may develop needs for social approval, influencing further his food preferences, and so on. Would this general approach to the understanding of the development of motivations be consistent with your ideas?

Would you say that basic, innate, instinctual patterns are modified by the environment or culture to which they are subjected?

DR. JUNG: Yes, certainly.

DR. EVANS: Also, concerning motivation, or the condition which arouses, directs, and sustains the individual, there appear to be two views found in much of our psychology in America today. One might be called an historical view, as illustrated by the biocentric theory just discussed, where we try to look at the history and development of the individual for answers as to why he is doing a certain thing at the moment.

Then we have another view, postulated and discussed by Dr. Kurt Lewin (1936) which is a field theory. He did not believe that the history—the past—was the most important element in motivation. Instead he suggested that all the conditions which affect the individual at a given moment help us to best understand the individual and predict his behavior. Do you think that the "present field" idea of Dr. Lewin has any virtue?

DR. JUNG: Well, obviously I always insist that even a chronic neurosis has its true cause in the moment—now. You see, the neurosis is made every day by the wrong attitude the individual has. On the other hand, however, that wrong attitude is a sort of fact that needs to be explained historically, by things that have happened in the past. But that is one-sided too, because all psychological facts are oriented, not only to a cause, but also to a certain goal. They are, in a way, teleological; namely, they serve a certain purpose, so the wrong attitude can have originated in a certain way long ago. It is equally true, however, that it wouldn't exist today anymore if there were not immediate causes and immediate purposes to keep it alive. Because of this, a neurosis can be finished suddenly on a certain day, despite all causes. One has observed, in the beginning of the war, cases of compulsion neurosis that have lasted for many years and suddenly were cured, because they got into an entirely new condition. It is like a shock, you see. Even the schizophrenic can be vastly improved by a shock because that's a new condition; it is a very shocking thing, so it shocks them out of their habitual attitude. Once they are no more in it, the whole thing collapses, the whole system that has been built up for years.

DR. EVANS: You have brought up many interesting and provocative ideas here. Another concept related to motivational development is the process of individuation, a process to which you frequently refer in your writing. Would you like to comment about this process of individuation, how all these factors move toward a whole—a totality?

DR. JUNG: Well, you know, that's something quite simple. Take an acorn, put it into the ground, and watch it grow and become an oak. That is man. Man develops from an egg, and develops into the whole man; that is the law that is in him.

DR. EVANS: So you think the psychic development is in many ways like the biological development.

DR. JUNG: The psychic development is out of the world; it is not something else, or an opinion. It is a fact that people develop in their psychical development on the same principle as they develop in the body. Why should we assume that it is a different principle? It is really the same kind of evolutionary behavior as the body shows. Consider, for instance, those animals that have specially differentiated anatomical characteristics, those of the teeth or something like that. Well, they have a mental behavior which is in accordance with those organs.

DR. EVANS: So as you see it, there is no need to bring in other types of ideas, other types of theories to explain development. The basic biological law is still . . .

DR. JUNG: The psyche is nothing different from the living being. It is the psychical aspect of the living being. It is even the psychical aspect of matter. It is a quality.

IV. ISSUES AND INSIGHTS

In these interviews Jung traces his pioneering efforts in the area of projective personality testing, specifically his word association test. He reacts positively to the general value of projective tests, although he indicates some reservations concerning the originality of Hermann Rorschach. Addressing himself to his work with patients, he discusses at some length the value of dream and fantasy material to the therapeutic process.

The reader will be interested in the strong feelings concerning mental telepathy which Jung possessed and his rather favorable reaction to the work of J. B. Rhine in this respect. He attempts to discuss his own complex contribution to understanding in this field by presenting his concept of synchronicity.

Jung's reaction to theories of psychosomatic medicine and tranquilizers alike reflect his reservations concerning how much real progress American medicine and American psychology have made. He cites his work with tubercular patients fifty years ago as evidence of his early understanding of psychosomatic medicine, for example, pointing out how slowly the American psychologist came to accept such a view.

This section concludes with Jung's comments on his contact with Einstein and Toynbee, and his feelings concerning the relative importance of statistics and a knowledge of the humanities for the beginning psychology student.

Diagnostic and Therapeutic Practices: Projective Tests, Rorschach, Word Association, Transference

DR. EVANS: We American psychologists do a great deal of testing, utilizing projective tests. As we discussed earlier, you certainly played a major role in developing projective testing with your word association method. What led you to develop the word association test?

DR. JUNG: You mean the practical use of it?

DR. EVANS: Yes.

DR. JUNG: Well you see, in the beginning when I was a young man, I was completely disoriented with patients. I didn't know where to begin or what to say; and the association experiment has given me access to their unconscious. I learned about the things they did not tell me, and I got a deep insight into things of which they were not aware. I discovered many things.

DR. EVANS: In other words, from such association responses you discovered complexes or areas of emotional blocks in the patient? Of course, the word "complex," which originated with you, is used very widely now.

DR. JUNG: Yes, complex—that is one of the terms which I originated.

DR. EVANS: Did you hope from these complexes or emotional blocks which you were uncovering as you administered this word association test to get at materials in the personal unconscious or the racial unconscious?

DR. JUNG: In the beginning there was no question of collective unconscious or anything like that. It was chiefly the ordinary personal complexes.

DR. EVANS: I see. You weren't expecting to get into such depth.

DR. JUNG: Among hundreds of complex associations, there might appear an archetypal element, but it wouldn't stand out particularly. That is not the point. You know, it is like the Rorschach, a superficial orientation.

DR. EVANS: You knew Hermann Rorschach, I believe, did you not?

DR. JUNG: No. He has circumvented me as much as possible.

DR. EVANS: But did you get to know him personally?

DR. JUNG: No. I never saw him.

DR. EVANS: In his terms, "introtensive" and "extrotensive," of course, he is reflecting your conceptions of introversion and extroversion, in my own estimation that is.

DR. JUNG: Yes, but I was anathema, because I was the one to first outline the concepts; and that, you know, is unforgivable. I never should have done it.

DR. EVANS: So you really didn't have any personal contacts with Rorschach?

DR. JUNG: No personal relations at all.

DR. EVANS: Are you familiar with Rorschach's test which uses inkblots?

DR. JUNG: Yes, but I never applied it, because later on I didn't even apply the word association test anymore. It just wasn't necessary. I learned what I had to learn from the exact examinations of psychic reactions; and that, I think is a very excellent means.

DR. EVANS: But would you recommend that other psychiatrists, clinical psychologists, and psychoanalysts use these projective tests, such as your word association test or Rorschach's test?

DR. JUNG: Well, perhaps. For the education of psychologists who intend to do actual work with people, I think it is the best means to make them see how the unconscious works.

DR. EVANS: So you feel that the projective tests have a function in training psychologists?

DR. JUNG: Yes. They are exceedingly didactic. With these tests one can actually demonstrate repression or the amnestic phenomenon, the way in which people cover their emotions, etc. It takes place like an ordinary conversation, but the tests provide certain principles and criteria which serve as guides and measuring devices for what one sees and hears.

It is all so interesting. You observe all the things you observe in a conversation with another person. For instance, in conversation when you ask a person something or begin to discuss certain things, you can observe certain things, little hesitations, mistakes in speech, etc.—all those things come to the fore. Then, what is more, in the experimental setting they are measurable.

I think I don't overrate the didactic value of projective tests. I think very highly of them in this capacity, that is, for educating young psychologists. And sometimes, of course, they are useful to any psychologist. If I have a case who doesn't want to talk, I can make an experiment and find out a lot of things through the experiment. I have, for instance, discovered a murder.

DR. EVANS: Is that right? Would you like to tell us how this was done?

DR. JUNG: You see, you have that lie detector in the United States, and that's like an association test I have worked out with the psycho-galvanic phenomenon. Also, we have done a lot of work on the pneumograph which will show the decrease of volume of breathing under the influence of a complex. You know, one of the reasons for tuberculosis is the manifestation of a complex. People have very shallow breathing, don't ventilate the apices of their lungs anymore, and get tuberculosis. Half of tuberculosis cases are psychic.

DR. EVANS: In working with a patient, would you say that it is essential for him to recapitulate his past life in order to help him deal with his present neurosis, as Freud did, or do you feel that you can deal situationally with his problem without going back and probing into things that happened to him during his early life?

DR. JUNG: There is no one-and-only system in therapy. In therapy you treat the patient as he is in the present moment, irrespective of causes and such things. That is all more or less theoretical. Sometimes I can start right away with posing the problem. There are cases who know just as much about their own neurosis as I know about it, in a way.

For instance, let us take the case of a professor of philosophy, an intelligent man, who imagines that he has cancer. He shows me several dozen X-ray plates that prove there is no cancer. He says, "Of course, I have no cancer, but nevertheless, I'm afraid I could have one. I've consulted many surgeons and they all assure me there is none; and I know there is none but I might have one." You see? That's enough. Such a case can stop from one moment to the next, just as soon as the person who has the sickness stops thinking such foolish things, but that is exactly what he cannot do.

In such a case, I say, "Well, it is perfectly plain to you that it is nonsense what you believe. Now why are you forced to believe such nonsense? What is the power that makes you think such a thing against your free will? You know it is all nonsense." It's like a possession. It is as though a demon were in him, making him think like that in spite of the fact that he doesn't want to. Then I say, "Now you have no answer; I have no answer. What are we going to do?" I add, "We will see what you dream for a starting point, because a dream is a manifestation of the unconscious side."

In this case our philosopher has never heard of the unconscious side, so I must explain to him about the existence of the unconscious; and I must explain to him that the dream is a manifestation of it. Thus, if we succeed in analyzing the dream, we may get an idea about that power, which is distorting his thinking. In such a case one can begin right away with the analysis of dreams, and the same is true for all cases that are a bit serious. Mind you, this is not a simple case, but a very difficult and serious case, in spite of the simplicity of the phenomenology, of the symptomatology.

In all cases after the preliminaries such as taking down the history of the family, the whole medical analysis, etc., we come to that question, "What is it in your unconscious that makes you wrong in your thinking, that hinders you from thinking normally?" Then we can begin with the observation of the unconscious, and the day-by-day process of analyzing the data produced by the unconscious. Now that we have discussed the first dream, the whole problem takes on new perspective, and he will have other dreams, each of which will have something to add until we have the whole picture. Now when we have the full picture, if he has the necessary moral stamina, he can be cured. In the end it is strictly a moral question, whether a man applies what he has learned or not.

DR. EVANS: Does your type approach, based on

introversion-extroversion constructs, help you in this analytical process?

DR. JUNG: Yes. I find in the study of the type, that it supplies a certain lead as to the personal nature of the unconscious, the personal quality of the unconscious in a given case. If you study an extrovert, you find that his unconscious has then an introverted quality. This is because all the extroverted qualities are played in consciousness, and the introverted qualities are all played in the unconscious; therefore, the unconscious has introverted qualities. The reverse composition, of course, is equally true. That knowledge gave me a lead of diagnostic value. It helped me to understand my patients. When I saw their conscious type, I got ideas as to their unconscious attitudes.

Now the neurotic is just as much controlled and influenced by the unconscious as he is by the conscious, so he may appear to be a type which actually is not a true diagnosis at all. In certain cases it is almost impossible to distinguish between conscious material and unconscious material, because you just cannot tell at first sight which is which. This has helped me to understand more the patients in terms of the Freudian emphasis (based on the past) as well as in Adlerian terms, which are more, as you say, concerned with the present situation of the patient.

In the course of years, I got quite a lot of empirical material about the peculiar way in which conscious and unconscious contents interact. I could do this by watching individuals who were actually going through analytical treatment. You see, there is a point when you try to integrate unconscious contents into consciousness; or you confront the patient who is holding a definite conscious attitude with the related unconscious attitude that is counteracting the conscious one. This process, of course, is perpetuating his neurosis; and it is just as though another personality of the opposite type were influencing him or disturbing him.

DR. EVANS: So, Professor Jung, you gradually developed through your typologies a sort of theory, a psychology of opposites, where the conscious revealed the qualities of one type and the unconscious revealed the qualities of the other type in a given individual. This would be a very important way, then, of helping you to analyze and understand the individual.

DR. JUNG: Yes, from a practical point of view, it is diagnostically quite important. The point I wanted to elucidate is that in analyzing a patient you create the expression of typical experiences during the therapeutic process. There is a sort of typical way in which the integration of consciousness takes place. The average way is that through the analysis of dreams, for instance, you become acquainted with the contents of the unconscious.

To begin with, you want to know all personal, subjective material about the individual, what sort of difficulties the individual has encountered in adapting to environmental conditions, etc. Now, it can be regularly observed that when you talk to an individual and this individual gives you insight into his inner preoccupations, interests, emotions, etc., or in other words, hands over his personal complexes, you get slowly and willy-nilly into a situation of a kind of authority. You are in possession of all the important items in a person's development, and you become a point of reference, because you are dealing with things which are very important to the person. I remember, for instance, that I analyzed a very well-known American politician, who told me any number of the secrets of his trade. Suddenly he jumped up and said, "My God, what have I done! You could get a million dollars for what I have told you now!" I said, "Well, I'm not interested. You can sleep in peace, because I shall not betray you. I'll forget it in a fortnight." So you see, that shows that the things people hand out are not merely indifferent things. When it comes to something emotionally important, they are handing out themselves. They are investing in the analyst big emotional value, just as if they were handing you a large sum of money or trusting you with the administration of their estate; they are entirely in your hands. Often I hear things that could ruin these people, utterly and permanently ruin them, things which would give me, if I had any blackmailing tendencies, unlimited power over them.

You can see that this kind of a situation creates an emotional relationship to the analyst, and this is what Freud called "transference," a central problem in analytic psychology. It is just as if these people had handed out their whole existence, and that can have very peculiar effects upon the individual. Either they hate you for it, or love you for it; but they are not indifferent. Thus, a sort of emotional relation between the patient and the doctor is fostered.

When a patient discusses such material, the content of it is associated with all the important persons in the life of that patient. Now the most important persons are usually father and mother in going back into a person's childhood. The first troubles are with the parents, as a rule. So, when a patient hands over to you his infantile memories about the father or mother, he also sees in you, the analyst, the image of that mother or father. Then it is just as if the doctor had taken the place of the father, or even of the mother. I have had quite a number of male patients that called me "Mother Jung," because they had handed over to me the image of their respective mothers, curiously enough. But you see, that's quite irrespective of the personality of the analyst. In this case, the personality of the analyst is simply disregarded. You now function as if you were the mother or father—the central authority. That is what one calls transference; that is projection.

Now, Freud doesn't exactly call it projection. He calls it transference, which is an allusion to an old, superstitious idea that if you have a disease, you can transfer the disease to an animal; or you can transfer the sin onto a scapegoat, and the scapegoat takes it out into the desert and makes it disappear. Thus, the patients hand over themselves in the hope that I can swallow that stuff and digest it for them. I am *in loco parentis* and I have a high authority. Naturally, I am also persecuted by the corresponding resistances, by all the manifold emotional reactions they have had against their parents.

Now that is the structure you have to work through first in analyzing the situation, because the patient in such a condition is not free; he is a slave.

He is actually dependent upon the doctor like a patient with an open abdomen on the operating table. He is in the hands of the surgeon, for better or for worse, so the thing must be finished. This means that we have to work through that condition in the hope that we will arrive at a different condition where the patient is able to see that I am not the father, not the mother, that I am an ordinary human being. Now everybody would assume that such a thing would be possible, that the patient could arrive at such an insight when he or she is not a complete idiot, that they could see that I am just a doctor and not that emotional figure of their fantasies. However, that is very often not the case.

I had a case once which involved an intelligent young woman, a student of philosophy who had a very good mind. One would easily think that she would be able to see that I was not her parental authority, but she was utterly unable to get out of this delusion. Now, in such a case one always has recourse to the dreams. She says through the conscious, "Of course, I know you are not my father, but I just feel like that. It is like you are my father; I depend upon you." Then I say, "Now we will see what the unconscious says." From that point, we work very hard in analyzing her dreams, and I begin to see that the unconscious is producing dreams in which I assume a very curious role.

In her dreams she is a little infant, sitting on my knee, and I am holding her in my arms. I have become a very tender father to the little girl, you know. More and more her dreams become emphatic in this respect; namely, I am a kind of giant, and she is a very little, frail human thing, quite a little girl in the hands of an enormous being. Then the final dream occurs in the series. In that dream, I was out in the midst of nature, standing in a field of wheat, an enormous field of wheat that was ripe for harvesting. I was a giant and I held her in my arms like a baby, with the wind blowing over the field.

Now as you know, when the wind is blowing over a wheat field, it waves; and with these waves I swayed, putting her to sleep. She felt as if she was in the arms of a god, of the "Godhead." I thought, "Now the harvest is ripe, and I must tell her," so I said, "You see, what you want and what you are projecting into me, because you are not conscious of it, is that unconsciously you are feeling the influence of a deity which does not possess your consciousness; therefore, you are seeing it in me." That clicked, because she had a rather intense religious education, that enabled her to understand. Of course, it all vanished later on and something disappeared from her world. The world became merely personal to her and a matter of immediate consciousness. That religious conception of the world was to her no longer existent, apparently. This makes sense, you see, because the idea of a deity is not an intellectual idea. It is an archetype, an archetypal idea, that catches hold of your unconsciousness, and once she could understand that consciously, the archetype could no longer control her.

You find this type of archetypal image practically everywhere under this or that name. You know, it has the name, "mana"—an all-powerful, extraordinary effect or quality—it doesn't matter whether it is personal at all or not. In the case of this girl, she suddenly became aware of an entirely heathenish image, an image that comes fresh from the archetype. She had not the idea of a Christian God, or of an Old Testament Yahweh, but of a heathenish God—a God of nature, a God of vegetation. He was the wheat itself. He was the spirit of the wheat, the spirit of the wind; and she was in the arms of that numen. Now that is the living experience of an archetype.

When that girl came to understand what was happening in her, it made a tremendous impression upon her. She saw what she really was missing, that missing value which she was projecting into me, making me indispensable to her. Then she came to see that I was not indispensable, because, as the dream says, she is in the arms of that archetypal idea. That is a numinous experience, you see, and that is the thing that people are looking for, an archetypal experience which is in itself an incorruptible value.

Until they have the experience and understand it, they depend upon other conditions; they depend upon their desires, their ambitions. They depend upon other people, because they have no value in themselves. They are only rational, and are not in possession of a treasure that would make them independent. Now when that girl could hold that experience, she no longer had to depend. The value became part of her. She had been liberated and was now complete. Inasmuch as she could realize such a numinous experience, she was and will be able to continue her part, her own way—her individuation. The acorn can become an oak, and not a donkey. Nature will take her course. A man or woman becomes that which he or she is from the beginning. I have seen quite a number of such cases as I have just cited to you.

DR. EVANS: How do the dreams and fantasies of the patient enter into this process?

DR. JUNG: I wrote a book about such dreams, you know, an introduction to the psychology of the unconscious. At that time my empirical material had been formed chiefly by observation of lunatics, cases of schizophrenia, and I had observed that there are, chiefly in the beginning of a disease, invasions of fantasies into conscious life, fantasies of an entirely unexpected sort which are most bewildering to the patient. He gets quite confused by these ideas, and he gets into a sort of panic since he never before has thought such things. They are quite strange to him and equally strange to his physician. Yes, the analyst is equally dumbfounded by the peculiar character of those fantasies. Therefore, one says, "That man is crazy. He is crazy to think such things; nobody thinks such things," and the patient agrees with him, which throws the patient into even more of a panic. So as an analyst I thought it to be really the task for psychiatry to elucidate that thing that broke into consciousness, the voices and the delusions. In those days, and mind you, I'm referring to over forty or fifty years ago, I had no hope to be able to treat these cases or to be able to help them, but I had a very great scientific curiosity which made me want to know what these things really were. You see, I felt that these things had a system and that they were not merely chaotic, decayed material, because there was too much sense in those fantasies.

This led me to begin studying cases of psychogenic diseases such as hysteria, somnambulism, and others where the content that flowed from the unconscious was in readable condition and capable of being understood. Then I saw that, in contradistinction to the schizophrenics, the mental contents of hysterics were elaborate, dramatic, suggestive, and insinuating, enabling one to make out a second personality. Now this is not the case in schizophrenia. There the fantasies, on the contrary, are unsystematic and chaotic, so that you cannot make out a second personality. The cases are of too complicated a nature. I needed a simpler type, or a more comprehensible type, to study.

An old professor of psychology and philosophy at the University of Geneva published a case concerning an American girl, wherein he described her half-poetic and half-romantic fantasies. He published that material without commenting on it, giving it as an example of creative imagination. Now, when I read those fantasies, I saw this as exactly the kind of material I needed. I was always a bit afraid to tell of my personal experiences with patients because I felt that people might say that too much suggestion was involved, but since I had no hand in this case, I could not be accused of having influenced the patient. That is the reason I analyzed those particular fantasies. That case became the object of a whole book called *The Psychology of the Unconscious*. I have revised it after forty years, and it is now called *Symbols of Transformation* (Jung, 1956).

In *The Psychology of the Unconscious,* I tried to show that there is a sort of unconscious that clearly produces things which are historical and not personal. At that time, I simply called it "the unconscious," not distinguishing between the two aspects involved. Using the fantasies of the American girl, I tried for the first time to produce a picture of the functioning of the unconscious, a functioning which pointed to certain conclusions as to the nature of the unconscious.

Writing that book cost me my friendship with Freud, because he couldn't accept it. To Freud, the unconscious was a product of consciousness, and the unconscious simply contained the remnants of consciousness; I mean that he saw the unconscious as a sort of storeroom where all the discarded things of consciousness were heaped up and left. To me, however, the unconscious was a matrix, a sort of basis of consciousness, possessing a creative nature and capable of autonomous acts, autonomous intrusions into the consciousness. In other words, I took the existence of the unconscious for a real fact, an autonomous factor that was capable of independent action.

To me that was a psychological problem of the very first order, and it made me think and feel, because the whole of philosophy, even up to the present day, has not recognized the fact that we have a counterfactor in our unconscious. It has not become recognized that in our psyche there are two factors, two independent factors, with consciousness representing one factor and, equally important, the unconscious representing the other factor. And the unconscious can interfere with consciousness any time it pleases. Now I say to myself, "This is very uncomfortable. I think I am the only master in my house, but in reality I must admit that there is another master, somebody in my house that can play tricks on me." I have to deal with the unfortunate victims of that interference every day in my patients.

I remember, for instance, one case which involved a learned and very rational man. He had a lot of personal problems, and these became so bad that he got into very disagreeable relations with his whole surroundings. He was a member of a society, and he got into a brawl with the people of that society; it was really quite shocking. He began having and reporting collective dreams to me. Suddenly, he dreamed of things he had never thought of in his life before, mythological motifs, and he thought he was crazy, because he could not understand it at all. It was just as if the whole world were suddenly transformed. You see this same process in a case of schizophrenia, but this was not a case of schizophrenia. In this case the collective dreams were expressing the mythological patterns or motifs which were in his unconscious.

There are many examples of this in the collective dreams I have published. To make it clear, I shall tell a long story. Then you will see how the collective dream applies in cases such as the one cited above. I have already mentioned the case of that intuitive girl who suddenly came out with the statement that she had a black snake in her belly. Well now, that is an example of a collective symbol. That is not an individual fantasy; it is a collective fantasy. That fantasy is well known in India. Now right at first, I even thought she might be crazy, for she had no more connection with India by all external considerations than I did. But, of course, we are all similar in at least one respect—we are all human. This girl was just highly intuitive and oriented toward a wholistic manner of thinking, or thinking always within a context of totality or wholeness, a mode of thinking which is known in and characteristic of India. It is the basis of a whole philosophical system, that of Tantrism, and this system has as its symbol Kundalini—Kundalini the serpent. This is something known only to some few specialists; it generally is not known that we have a serpent in the abdomen. Well, that is a collective dream or collective fantasy.

DR. EVANS: As the individual goes through life day-to-day, is it possible that things that trouble him and cause tension lead to repression?

DR. JUNG: He doesn't repress consciously always. These things disappear, and Freud explains that by active repression. But you can prove that these things never have been conscious before. They simply don't appear, and you don't know why they don't appear. Of course, when they do come up later, one can give the explanation that they have not appeared before because they were in disagreement or were incompatible with the patient's conscious views and attitudes. But that is afterwards that you can say this; you were not able to predict it.

So you see, these things that have an emotional tone are partially autonomous. They can appear or not appear. They can disappear at will, not of the subject, but of their own; and, also, you can repress them. It is just so the same as with projections. For instance, people say, "One makes projections." That's nonsense. One doesn't make them; one finds them. They are already there; they are already in the unconscious. And so, these disappearances, or the so-called repressions, are just like projections. Without your having anything to do with it, they are already part of the unconscious. There are cases, sure, where consciousness enters in, but I should say that the majority of cases are unconscious. That was my first point of difference with Freud. I saw in the association experiment that certain complexes are quite certainly not repressed. They simply won't appear. This is because, you see, the unconscious is real; it is an entity; it works by itself; it is autonomous.

DR. EVANS: So in a sense, looking at the so-called defense mechanisms, projection, rationalization, etc., you would differ from the orthodox psychoanalytic view in that you would not say that they are developed as a means of defending the ego. Rather, you would say that they are already there as manifestations of patterns that are already present in the unconscious.

DR. JUNG: Yes. Take, for instance, the example of that serpent. That never had been repressed, or else it would have been conscious to her. On the contrary, it was unconscious to her and only appeared in her fantasies. It appeared spontaneously. She didn't know how she came to it. She said, "Well, I just saw it."

DR. EVANS: Now some of the orthodox psychoanalysts might have said, "This is a phallic symbol."

DR. JUNG: But you can say anything, you know. One can say that a church spire is a phallic symbol, but what is it when you dream of a penis? You know what an analyst said, one of the orthodox men, one of the old guard? His explanation of that question was that in this case the censor had not functioned. You call that a scientific explanation?

Psychic Phenomena and Drugs

DR. EVANS: You are, familiar, of course, with the work of Dr. J. B. Rhine at Duke University. Some of his work in extrasensory perception and clairvoyance, or mental telepathy, sounds much like the research into intuitive function, a phase of your work which we discussed earlier. For example, would you say that a person who has clairvoyance would be an intuitive type in your frame of reference?

DR. JUNG: That's quite probable. Or it can be a sensation type, say an extrovert-sensation type who is very much influenced by the unconscious. He has introverted intuition in his unconscious.

139

DR. EVANS: Dr. Jung, you speak of rational and irrational functions, thinking and feeling being rational, and sensation and intuition being irrational. Would you care to elaborate on this notion?

DR. JUNG: As you say, there are two groups, the rational group and the irrational group. The rational group consists of the two functions, thinking and feeling. The ideal thinking is a rational result. Feeling is also a rational result. They hold rational values. That is differentiated thinking.

The irrational group is comprised of sensation and intuition. Sensation functions in such a way that it may not prejudice facts; it shall not prejudice facts. To the sensation type, the ideal perception is that you have an accurate perception of things as they are without additions or corrections. On the other side, intuition does not look at things as they are. That is anathema to the intuition. It looks ever so shortly at things as they are, and makes off into an unconscious process at the end in which he will see something nobody else will see.

DR. EVANS: So in terms of the person who is clairvoyant . . .

DR. JUNG: Those people who yield the best results are always those people who are introverted or where introverted intuition comes in. But that is a side aspect of it; it is not interesting.

There is another question far more interesting, namely, the terms they use. Rhine himself uses them —recognition, telepathy, etc. They mean nothing at all. They are words, but he thinks he has said something when he says "telepathy."

DR. EVANS: The word itself is not a description of the process.

DR. JUNG: Not a description. It means nothing, nothing at all.

DR. EVANS: Now, of course, a lot of the things that you are describing, some scientists would insist are due to chance, chance occurrences, and chance factors. In his own work, Rhine used statistical probability analysis methods. He reports these occurrences more often than would be expected by chance.

DR. JUNG: Well you see, he proves that it is more than chance; it is statistically plausible. That is the important point which hasn't been contradicted.

There was some experimental proof offered in England, which resulted in the accusation: "Oh, Rhine, that's nothing but guesswork." And that is exactly true; that is guessing, what you call guessing. However, a hunch is guessing, but a definite guess, you know. All this really means nothing.

You see, the point is that it is more than merely probable; it is beyond chance. That's the major point. But as you know, people hate such problems they can't deal with concretely, and they can't deal with this one concretely. In fact, even Rhine does not understand how often extrasensory phenomena really occur, because it is a relativation. Now I am going to say something which in these sacred rooms is anathema, a relativation of time and space through the psyche. That's the fact; that is what Rhine has made evident, but for scientists to say, "I'll swallow that," now that is difficult.

DR. EVANS: We might go a little further into some

of your recent works in this area which many consider quite profound, but are not too well-known to many of our students.

DR. JUNG: Of course not. Nobody reads these things, only the general public, because my books are at least sold.

DR. EVANS: To be more specific, I'm referring to the concept, synchronicity, which you have discussed, and which has some relevance at this point in our discussion. Would you care to comment on synchonicity?

DR. JUNG: That is awfully complicated. One wouldn't know where to begin. Of course, this kind of thinking started long ago, and when Rhine brought out his results, I thought, "Now we have at least a more or less dependable basis to argue on." But the argument has not been understood at all, because it is really very difficult.

When you observe the unconscious, you will come across plenty of cases which show a very peculiar kind of parallel events. For example, I have a certain thought of a certain definite subject which is occupying my attention and my interest; and at the same time something else happens, quite independently, that portrays just that thought. This is utter nonsense, you know, looked at from a causal point of view. However, that there is something else to it which is not nonsense is made evident by the results of Rhine's experiments. There is a probability; it is something more than chance that such a case occurs.

I never made statistical experiments except one in the way of Rhine. I made one for another purpose.

But I have come across quite a number of cases where it was most astounding to find that two causal chains happened at the same time, but independent of each other, so that you could say they had nothing to do with each other. It's really quite clear. For instance, I speak of a red car and at that moment a red car comes here. Now I haven't seen the red car, because it wasn't possible; it was hidden behind the building until just this moment when it suddenly appeared. Now many would say that this is an example of mere chance, but the Rhine experiment proves that these cases are not mere chance.

Now it would be superstitious and false to say, "This car has appeared because here were some remarks made about a red car; it is a miracle that a red car has appeared." It is not a miracle; it is just chance—but these chances happen more often than chance allows. That shows that there is something behind it.

Rhine has a whole institute, many co-workers, and has the means. We have no means here to make such experiments; otherwise, I probably would have done them. Here it is just physically impossible, so I have to content myself with the observation of facts!

DR. EVANS: An interesting area which is being discussed a lot in the United States today, and I'm sure is of interest to you as well, is that of psychosomatic medicine, an area dealing with the way in which emotional components of personality can affect bodily functions.

DR. JUNG: As an example of this, I see a lot of astounding cures of tuberculosis—chronic tuberculosis

—effected by analysts; people learn to breathe again. The understanding of what their complexes were— that has helped them.

DR. EVANS: When did you first become interested in the psychic factors of tuberculosis? Many years ago?

DR. JUNG: I was an analyst to begin with; I was always interested naturally. Maybe also because I understood so little of it, or more importantly, I *noticed* that they understood so little.

DR. EVANS: To expand on my earlier question, we are right now becoming more and more interested in the United States in how emotional, unconscious personality factors can actually have an affect on the body. Of course, the classic example in the literature is the peptic ulcer. It is believed that this is a case where emotional factors have actually created pathology.

These ideas have been extended into many other areas. It is felt, for example, that where there already is pathology, these emotional factors can intensify it. Or sometimes there may be actual symptoms or fears concerning pathology when no true pathology exists, such as in cases of hysteria or hypochondriasis. For example, many physicians in America say that sixty to seventy percent of their patients do not have anything really physically wrong with them, but they instead have disorders of psychosomatic origin.

DR. JUNG: Yes, that is well-known—since more than fifty years. The question is how to cure them.

DR. EVANS: Speaking of such psychosomatic disturbances, as, for instance, your experiences and studies into tuberculosis, do you have any ideas as to why the patient selects this type of symptom?

DR. JUNG: He doesn't select; they happen to him. You could ask just as well when you are eaten by a crocodile, "How did you happen to select that crocodile?" Nonsense, he has selected you.

DR. EVANS: Of course, "selected" in this sense refers to an unconscious process.

DR. JUNG: No, not even unconsciously. That is an extraordinary exaggeration of the importance of the subject, to say he was choosing such things. They get him.

DR. EVANS: Perhaps one of the most radical suggestions in the area of psychosomatic medicine has been the suggestion that some forms of cancer may have psychosomatic components as causal factors. Would this surprise you?

DR. JUNG: Not at all. We know these since long ago, you know. Fifty years ago we already had these cases; ulcer of the stomach, tuberculosis, chronic arthritis, skin diseases. All are psychogenic under certain conditions.

DR. EVANS: And even cancer?

DR. JUNG: Well you see, I couldn't swear, but I have seen cases where I thought or wondered whether or not there was a psychogenic reason for that particular ailment; it came too conveniently.

Many things can be found out about cancer, I'm sure. You see, with us it has been always a question of how to treat these things, because any disease possible has a psychological accompaniment. It just all depends upon—perhaps life depends upon it—whether you treat such a patient psychologically in the proper way or not. That can help tremendously, even if you cannot prove in the least that the disease in itself is psychogenic.

You can have an infectious disease in a certain moment, that is, a physical ailment or predicament, because you are particularly accessible to an infection —maybe sometimes because of a psychological attitude. Sore throat is such a typical psychological disease; yet it is not psychological in its physical consequences. It's just an infection. So you ask, "Then why does psychology have anything to do with it?" Because it was the psychological moment maybe that allowed the infection to grow. When the disease has been established and there is a high fever and an abscess, you cannot cure it by psychology. Yet it is quite possible that you can avoid it by a proper psychological attitude.

DR. EVANS: So all this interest in psychosomatic medicine is pretty old stuff to you.

DR. JUNG: These things were all known here long ago.

DR. EVANS: And you are not at all surprised at the new developments?

DR. JUNG: No. For instance, there is the toxic aspect of schizophrenia. I published it fifty years ago—just fifty years ago—and now everyone discovers it. You are far ahead in America with technological things, but in psychological matters and such things, you are fifty years back. You simply don't understand them; that's a fact. I don't want to offend you; that's a general corrective statement; you simply are not yet aware of what there is. There are plenty more things than people have any idea of. I told you that case of the philosopher who didn't even know what the unconscious was; he thought it was an apparition. Everyone who says that I am a mystic is just an idiot. He just doesn't understand the first word of psychology.

DR. EVANS: There is certainly nothing mystical about the statements you have just been making. Now to pursue this further, another development that falls right in line with this whole discussion of psychosomatic medicine has been the use of drugs to deal with psychological problems. Of course, historically drugs have been used a great deal by people to try to forget their troubles, to relieve pain, etc. However, a particular development has been the so-called nonaddictive tranquilizing drugs. These, of course, became prominent in France with the drug, chlorpromazine. Then followed such drugs as reserpine-serpentina, and a great variety of milder tranquilizers, known by such trade names as Miltown and Equinal. They are now being administered very freely to patients by general practitioners and internists. In other words, not only are the stronger tranquilizers being administered to mentally ill patients such as schizophrenics, but to a great extent today these drugs are being dispensed almost as freely as aspirins to reduce everyday tensions.

DR. JUNG: This practice is very dangerous.

DR. EVANS: Why do you think this is dangerous? These drugs are supposed to be nonaddictive.

DR. JUNG: It's just like the compulsion that is

caused by morphine or heroin. It becomes a habit. You don't know what you do, you see, when you use such drugs. It is like the abuse of narcotics.

DR. EVANS: But the argument is that these are not habit-forming; they are not physiologically addictive.

DR. JUNG: Oh yes, that's what one says.

DR. EVANS: But you feel that psychologically there is still addiction?

DR. JUNG: Yes. For instance, there are many drugs that don't produce habits, the kind of habits that morphine does; yet it becomes a different kind of habit, a psychical habit, and that is just as bad as anything else.

DR. EVANS: Have you actually seen any patients or had any contact with individuals who have been taking these particular drugs, these tranquilizers?

DR. JUNG: I can't say. You see, with us there are very few. In America there are all the little powders and the tablets. Happily enough, we are not yet so far. You see, American life is in a subtle way so one-sided and so uprooted that you must have something with which to compensate the real nature of man. You have to pacify your unconscious all along the line because it is in absolute uproar; so at the slightest provocation you have a big moral rebellion in America. Look at the rebellion of modern youth in America, the sexual rebellion, and all that. These rebellions occur because the real, natural man is just in open rebellion against the utterly inhuman form of American life. Americans are absolutely divorced from nature in a way, and that accounts for that drug abuse.

DR. EVANS: But what about the treatment of individuals who are seriously mentally ill? We have the problem of hospitalized, psychotic patients. For instance, certain schizophrenics are so withdrawn that they are virtually impossible to interact with in psychotherapy; so in many hospitals in the United States, drugs such as chlorpromazine have been used in order to render many such patients more amenable to psychotherapy. I don't think most of our practitioners believe the drugs cure the patients in themselves, but they at least make the patient more amenable to therapy.

DR. JUNG: Yes, the only question is whether that amenability is a real thing or drug-induced. I am sure that any kind of suggestive treatment will have effect, because these people simply become suggestible. You see, any drug or shock in the mind will lower stamina, making these people accessible to suggestion. Then, of course, they can be led, can be made into something, but it is not a very happy result.

On Scientific "Truth"

DR. EVANS: To change the topic for a moment, Professor Jung, I know our students would be interested in your opinion concerning the kind of training and background a psychologist, a person who wants to study the individual, should have. For example, there is one view that says maybe he should be trained primarily as a rigorous scientist, a master of such tools as statistics and experimental design. Others feel, however, that a study of the humanities is also important for the student who wants to study the individual.

DR. JUNG: Well of course, when you study human

psychology, you can't help noticing that man's psychology doesn't only consist of the ramifications of instinct in his behavior. There are other determinants, many others, and the study of man from his biological aspect only is by far insufficient. To understand human psychology, it is absolutely necessary that you study man also in his social and general environments. You have to consider, for instance, the fact that there are different kinds of societies, different kinds of nations, different traditions; and in the interest of that purpose, it is absolutely necessary that one treat the problem of the human psyche from many standpoints. Each is naturally a considerable task.

Thus, after my association experiments at which time I realized that there was obviously an unconscious, the question became, "Now what is this unconscious? Does it consist merely of remnants of conscious activities, or are there things that are practically forever unconscious? In other words, is the unconsicous a factor in itself?" And I soon came to the conclusion that the unconscious must be a factor in itself. You see, I observe time and again, for instance, when delving into people's dreams or schizophrenic patients' delusions and fantasies, that therein is contained motives which they couldn't possibly have acquired in our surroundings. This, of course, depends upon the belief that the child is not born *tabula rasa*, but instead is a definite mixture or combination of genes; and although the genes seem to contain chiefly dynamic factors and predispositions to certain types of behavior, they have a tremendous importance also for the arrangement of the psyche, inasmuch as it appears naturally. Before you can see into the psyche, you cannot study it, but once it appears, you see that it has certain qualities and a certain character. Now the explanation for this must depend upon the elements born in the child, so factors determining human behavior are born within the child, and determine further development. Now that is one side of the picture.

The other side of the picture is that the individual lives in connection with others in certain definite surroundings that will influence the given combination of qualities. And that now is also a very complicated factor, because the environmental influences are not merely personal. There are any number of objective factors. The general social conditions, laws, convictions, ways of looking at things, of dealing with things; these things are not of an arbitrary character. They are historical. There are historical reasons why things are as they are. There are historical reasons for the qualities of the psyche and there is such a thing as the history of man's evolution in past aeons, which as a combination show that real understanding of the psyche must consist in the elucidation of the history of the human race—history of the mind, for instance, as in the biological data. When I wrote my first book concerning the psychology of the unconscious, I already had formed a certain idea of the nature of the unconscious. To me it was then a living remnant of the original history of man, man living in his surroundings. It is a very complicated picture.

So you see, man is not complete when he lives in a world of statistical truth. He must live in a world where the "whole" of man, his entire history, is the concern; and that is not merely statistics. It is the expression of what man really is, and what he feels himself to be.

The scientist is always looking for an average. Our natural science makes everything an average, reduces everything to an average; yet the truth is that the carriers of life are individuals, not average numbers. When everything is statistical, all individual qualities are wiped out, and that, of course, is quite unbecoming. In fact, it is unhygienic, because if you wipe out the mythology of a man, his entire historical sequence, he becomes a statistical average, a number; that is, he becomes nothing. He is deprived of his specific value, of experiencing his own unique value.

You see, the trouble is that nobody understands these things apparently. It seems quite strange to me that one doesn't see what an education without the humanities is doing to man. He loses his connection with his family, his connection with his whole past—the whole stem, the tribe—that past in which man has always lived. We think that we are born today *tabula rasa* without a history, but man has always lived in the myth. To think that man is born without a history within himself—that is a disease. It is absolutely abnormal, because man is not born every day. He is born into a specific historical setting with specific historical qualities, and therefore, he is only complete when he has a relation to these things. If you are growing up with no connection from the past, it is like being born without eyes and ears and trying to perceive the external world with accuracy. Natural science may say, "You need no connection with the past; you can wipe it out," but that is a mutilation of the human being. Now I saw from a practical experience that this kind of proceeding has a most extraordinary therapeutic effect. I can tell you such a case.

There was a Jewish girl. Her father was a banker. She had been educated more through worldly experience than formal education, and was decidedly lacking in any understanding of tradition. I examined her history further and found out that her grandfather had been a Saddik in Galicia. With this insight, I knew the whole story, and let me explain why. This particular girl suffered from phobia, a terrific phobia, and had been under psychoanalytic treatment already with no effect. She was really badly plagued by that phobia, in excited states and so on. I observed that this girl had blocked significant influences of her past. For instance, the fact that her grandfather was a Saddik— he lived in the myth, was one influence she had blocked. Her father too had resisted this influence. So I simply told her, "You will stand up to your fears if you gain insight into what you have lost or are resisting. Your fear is the fear of the influences frcm the past." You know, the effect was that within a week she was cured from so many years of bad anxiety states, because this insight went through her like a lightning bolt. I was able to interpret the source of the problem so quickly because I knew that she was absolutely lost. She thought she was in the middle of things, functioning well, but actually she was, in a sense, lost or gone.

DR. EVANS: What can we learn from this remarkable case, Dr. Jung?

DR. JUNG: Well, it illustrates that it makes no sense and that our existence is incomplete when we are just "average numbers." The more you make people into average numbers, the more you destroy our society. The ideal state and the slave state come into being. If you want to be an "average number," go to Russia. There it is wonderful; there you can be a number. But one pays very dearly; our whole life goes to blazes, like in the case of the girl. I have plenty of cases of a similar kind.

DR. EVANS: As one reads your work, we seem to be aware that you know archeology, anthropology . . .

DR. JUNG: Well, this is true inasmuch as a great deal of my work is concerned with these disciplines, but I have no mathematical gifts, you know, which handicaps me some. You cannot get real knowledge or understanding of nuclear physics without a good mastery of mathematics, higher mathematics. There I only have a certain relation with it on the epistemological questions, you know. Modern physics is truly entering the sphere of the invisible and intangible, as it were. It is in reality a field of probabilities, which is exactly the same as the unconscious. I often have discussed this with Professor Pauli (1955). Now he is a nuclear physicist, and to my amazement I found that they have terms which are used in psychology too. This is simply on account of the fact that we are both entering a sphere which is unknown. The physicist enters it from without and the psychologist from within. That's the reason for the parleys between psychology and higher mathematics. For instance, we psychologists use the term "transcendental function." Now transcendental function is a mathematical concept, the function of rational and imaginary numbers. Now that is higher mathematics, with which I have nothing to do. But we come to the same terminology.

Einstein
and
Toynbee

DR. EVANS: When you spoke with Dr. Einstein in your early discussion, he more or less tried out some of his ideas on you. Did you ever bring to him the possibility that relativity might apply to psychic functions? Did you ever discuss that?

DR. JUNG: Well, you see, you know how it is when a man is so concentrated upon his own ideas as was Dr. Einstein; and when he is a mathematician on top of everything, then you are not welcome.

DR. EVANS: What year was it that you were friends with Einstein?

DR. JUNG: I wouldn't call myself a friend. I was

simply the host. I tried to listen and to understand, so there was little chance to insert some of my own ideas.

DR. EVANS: Was this after he had already formulated his relativity theories, or just before?

DR. JUNG: That was when he was working on it, right in the beginning. It was very interesting.

DR. EVANS: In your dealings with Professor Toynbee, have you gotten rather interested in his ideas of history?

DR. JUNG: Ah, yes, particularly his ideas about the life cycles of civilizations, and the way that they are ruled by archetypal forms. Toynbee has seen what I mean by historical functions of archetypal developments. That is a mighty important determinant of human behavior, and can span centuries or thousands of years. It expresses itself in symbols, sometimes symbols that you would never think of at all. For instance, as you know, Russia, the Soviet Republic, had that symbol of the red star. Now it is a five-rayed red star. America has the five-rayed white star. They are enemies; they can't come together. In the Middle Ages for at least 200 years the red and white were the couple; they were ultimately destined to marry each other. Now America is a sort of matriarchy, inasmuch as most of the money is in the hands of women, and Russia is the land of the little father; it's a patriarchy. So the two are mother and father. To use the terminology of the Middle Ages, they are the white woman, the *femina candida,* and the red slave, the *servus rubeus.* The two lovers have quarreled with each other.

DR. EVANS: Well, Dr. Jung, you've patiently and interestingly responded in a spontaneous fashion to questions ranging from your feelings about Freud's ideas to reactions to Toynbee. Perhaps we should not impose upon your extreme kindness any longer at this time. I do hope, however, that our students are stimulated by what you've said to go back to your great array of writings. After all, this is the real purpose for making these interviews available to students, to motivate them to read the original writings of the world's great contributors to our understanding of man's personality.

DR. JUNG: Yes. People have to read the books, by golly, in spite of the fact that they are thick. I'm sorry.

V. CONCLUSION

As we completed our discussions, I attempted to bring certain themes into focus; for example, Jung's main points of difference with Freud. His reactions to the questions concerning specifics in Freudian theory often reflected a form of tolerance—suggesting that Freud hadn't gone far enough or was dealing with something obvious—rather than clear disagreement. To Jung, the Oedipus complex was but one of an infinite number of archetypes. He seemed to be poking fun at Freud for "discovering" this one archetype and assuming that the whole of mankind pivoted around it, disregarding the other archetypes. He also seemed to feel that many of the details of psychosexual development as described by Freud were rather asinine. There was a definite note of sarcasm accompanying his discussion of the questions concerning Freud's oral and anal levels of development. He did not dismiss these ideas; he suggested that they were so self-evident as to be unimportant.

Another interesting point of disagreement centered around the importance of the power motive. Jung contended that Nietzsche's power concepts, although consciously ignored by Freud, were nonetheless masked in terms of sex in Freud's theories. Furthermore, he seemed to see the libido as having an important power component, as well as encompassing other needs of the individual besides sex. He insisted on viewing Freud's conception of the libido as being centered on sexual energy.

It is unlikely that Freud ever intended to be interpreted as literally as Jung often did during the course of the interviews. I may have set the stage for such literal interpretations on the part of Jung because of the need to make the questions both clear and provocative, though Jung virtually never challenged my often incomplete—as well as literal—descriptions of Freud's conceptions.

In my efforts to arrive at the essences of Jung's theory of personality, I was delighted with his disarming way of admitting that one of his ideas was complex or not easily understood. For example, I had found Jung's conceptualizations in describing his intuitive function rather vague. In response to a question concerning the intuitive-introverted type, Jung readily admitted that it was difficult to explain. He then proceeded to give a lengthy description involving some remarkable case histories, such as the one with the girl who believed she had a snake in her abdomen. He accepted honestly and without pretension this task of explanation and the necessity to clarify.

He made very sure that we understood that not only did he "invent" the commonly known terms, "introvert" and "extrovert," but the term "complex," as well. But in this instance, as in several others, the twinkle in his eyes made it clear that he was not engaging in blatant braggadocio, but was just trying to set the record straight.

He responded openly to my questions regarding alleged Nazi sympathies, and even accusations of anti-Semitism. He vehemently denied the accusations, and replied to my question by saying that to him, when Freud and the others fled the Third Reich, there seemed to be great danger that the power of the psychoanalytic movement would diminish, since Germany had been so central to it. He assumed the editorship of the Berlin *Psychoanalytic Journal* as a means of maintaining at least this scholarly center of the movement. He remained puzzled as to why this was construed as being evidence of sympathy for the Nazis. He admitted that Hitler was a phenomenon to be studied, but had only complete contempt for what he represented and what he did in every respect. He cited a number of experiences to illustrate the nature of his relationship with Jewish individuals throughout his life and indicated how Jews who had known him most intimately, some of whom had been among the more significant developers of many of his ideas, had taken it upon themselves to contradict the myth of his anti-Semitism. He concluded by asking how anyone could truly understand the breadth of his theories concerned with understanding the individual and accuse him of being prejudiced toward believers in a religion which has reflected the wisdom of the ages.

Another area that seemed to trouble him was the accusation that he was a mystic. Notwithstanding the fact that much of his writing, aside from his formulations concerning personality which are reflected in the present volume, tends to deal with metaphysical, transcendental and frankly spiritual problems, Jung appeared to be troubled by the label mystic. It was as if Jung the physician, psychologist, and scientist had long been engaged in battle with Jung the philosopher and speculator. It would seem that when Jung objected to being designated a mystic, he was suggesting that higher, more complex forms of scientific inquiry will some day validate even his least concrete conceptualizations. He seemed to feel that a creative breakthrough might alter the course of what we believe the true domain of science to be.

Perhaps his own summary of his goals is the best conclusion we can draw, as he states in the autobiographical *Memories, Dreams and Reflections*:

> I myself am haunted by the same dream, and from my eleventh year I have been launched upon a single enterprise which is my "main business." My life has been permeated and held together by one idea and one goal: namely, to penetrate into the secret of the personality. Everything can be explained from this central point, and all my works relate to this one theme.

APPENDIX: COMPLETE TRANSCRIPTS OF THE FOUR ONE-HOUR DISCUSSIONS BETWEEN C. G. JUNG AND RICHARD I. EVANS, AUGUST 5-8, 1957

This section contains our original transcriptions from the soundtrack of the four one-hour filmed discussions between C. G. Jung and Richard I. Evans, on which the edited version in this volume is based. The strict time limits and the amount of material to be included left little leeway for editing, so that most of the material contained in this section appears in a more ordered form in Chapters I—IV. These transcripts, further edited by William McGuire, are to be included in a forthcoming volume of the *Collected Works of C. G. Jung* and will be of interest to the Jungian scholar and to those readers who are interested particularly in the dialogue style.

First
Interview

Text pages 41–49

DR. EVANS: Dr. Jung, many of us who have read a great deal of your work are aware of the fact that in your early work you were to some degree at least in association with Dr. Sigmund Freud, and I know it would be of great interest to many of us to hear a little bit about how you happened to hear of Dr. Freud and how you happened to become involved with some of his work and ideas.

DR. JUNG: Well, as a matter of fact, it was in the year 1900, in December, soon after Freud's book about dream interpretation had come out, that I was asked by my chief, Professor Bleuler, to give a review of the book. I studied the book very attentively and I didn't understand many things in it, which were not clear at all, to me. But from other parts I got the impression that this man really knew what he was talking about, and I thought that this is certainly a masterpiece—full

169

of future. I had no ideas then of my own; I was just beginning. It was just when I began my career as assistant in the psychiatric clinic. And then I began with experimental psychology, or psychopathology; I applied the experimental association methods of Wundt and the same that has been applied at [Kraepelin's] psychiatric clinic in Munich, and I had studied the results and had the idea that one should go once more over it, so I made these association tests and I found out that the important thing in them had been missed, because it is not interesting to see that there is a reaction—a certain reaction—to a stimulus word; that is more or less uninteresting. But the interesting thing is—why people could *not* react to certain stimulus words, or in any entirely inadequate way. And then I began to study these places in the experiment where the attention, or the capability apparently of the test person began to waver or to disappear, and I found it out soon that it is a matter of intimate personal affairs people were thinking of, or which were in them, even if they momentarily did not think of them, when they were unconscious, in other words; that nevertheless the inhibition came from the unconscious and hindered the expression in speech. Then, in examining all these cases as carefully as possible, I saw that it was a matter of what Freud called repressions. I also saw what he meant by symbolization.

DR. EVANS: In other words, from your word association studies, some of the things in *The Interpretation of Dreams* began to fall into place?

DR. JUNG: And then I wrote a book about the psychology of dementia praecox, it was called then—now it is schizophrenia—and I sent the book to Freud and wrote to him about my association experiments and how they confirmed his theory thus far. That is how my friendship with Freud began.

DR. EVANS: There are other individuals who also became interested in Dr. Freud's work, and one of them was Dr. Alfred Adler. As you remember Dr. Adler, what was your impression of the thing that led him to become interested in Dr. Freud's work?

DR. JUNG: He belonged; he was one of the young doctors that belonged to his surroundings there. There were about twenty young doctors who followed Freud there, who were—who had a sort of little society, and Adler was one who happened to be there and he learned, he studied Freud's psychology in that circle.

DR. EVANS: Another individual, of course, who joined this group was Otto Rank, and of course he, unlike yourself, Dr. Adler and Dr. Freud, was not a physician; did not have the Doctor of Medicine degree. Was this regarded by your group at this time as something unusual to have someone become interested in these ideas who was not by training a physician?

DR. JUNG: Oh, no. I have met many people who represented different faculties that were interested in psychology. All people who have to do with human beings were naturally interested; theologians, lawyers, pedagogues; they all have to do with the human mind and these people were naturally interested. I'm naturally prejudiced, you know.

DR. EVANS: Then your group, including Freud, did not feel that this was exclusively an area of interest for the physician? This was something that might appeal to many?

DR. JUNG: Oh my, yes! Mind you, every patient you have gets interested in psychology, inevitably. Nearly everyone thinks he is meant to be an analyst.

DR. EVANS: One of the very fundamental ideas of the original psychoanalytic theory was Freud's conception of the libido as a sort of broad, psychic sexual energy. Of course we all know that you began to feel that Dr. Freud might have laid, perhaps, a little bit too much stress on sexuality in his theories. When did you begin first feeling this?

DR. JUNG: In the beginning I had naturally certain prejudices against these conceptions, and after a while I overcame them. I could do that owing to my biological training; I could not deny the importance of the sexual instinct, you know, but later on I saw that it was really one-sided because you see man is not only governed by the sex instinct; there are other instincts as well. For instance, in biology you see the nutritional instinct is just as important as the sex instinct, although in primitive societies sexuality plays a role much smaller than food. Food is the all-important interest and desire. Sex—that is something they can have everywhere—they are not shy. But food is difficult to obtain, you see, and so it is the main interest. Then in other societies, for instance, I mean civilized societies, the power drive plays a much greater role than sex. For instance, there are many big-business men who are impotent because their whole energy is going into moneymaking or dictating the laws to everybody else, that is much more interesting than affairs with women.

DR. EVANS: So in a sense, as you began to look over Dr. Freud's emphasis on sexual drive, you began to think in terms of other cultures—broader ideas—and it began to seem to you that this wasn't really an emphasis that was sufficiently universal to be important?

DR. JUNG: Well, you know, I couldn't help seeing it, because I had studied Nietzsche. I knew the work of Nietzsche very well; he had been a professor at Basel University, and the air was full of talk about Nietzsche, and so naturally I had studied his works, and there I saw an entirely different psychology, which was also a psychology—a perfectly competent psychology, but all built upon the power drive.

DR. EVANS: Do you think it possible that Dr. Freud was either ignoring Nietzsche, or had perhaps not wanted to be influenced by Nietzsche?

DR. JUNG: You mean his personal motivation?

DR. EVANS: Yes.

DR. JUNG: Of course it was a personal prejudice. It happened to be his main point, you know, that certain people are chiefly looking for this side and other people to another side. So, you see, the inferior Dr. Adler—the younger, the weaker, naturally had a power complex. He wanted to be the successful man; Freud was a successful man; he was on top, and so he was interested only in pleasure and the pleasure principle, and Adler was interested in the power drive.

DR. EVANS: You feel that it was a sort of function of his own personality?

DR. JUNG: Yes—it is quite natural; it is one of two ways of how to deal with reality. Either you make reality an object of pleasure if you are powerful enough already, or you make it an object of your desire to grab it, or possess it.

DR. EVANS: Some observers have thought that perhaps the patients that Dr. Freud saw in Vienna of this period were often individuals who were repressed sexually, and perhaps so many of his patients having been of this type, this may have been one thing that sort of reinforced Dr. Freud's ideas. In other words, in a sense, that the Viennese society was a rather "repressed" society. Do you think this might have been a factor?

DR. JUNG: Well, it is certainly so that in the end of the Victorian age there was a reaction going over the whole world against the sex taboos, so-called. One didn't understand anymore properly why or why not; and Freud belongs in that time; a sort of liberation of the mind of such taboos.

DR. EVANS: There was a reaction, then, against the sort of tight, inhibited culture he was living in.

DR. JUNG: Yes; Freud in that way, on that side, really belonged to the category of a Nietzschean mind. Nietzsche had liberated Europe from a great deal of such prejudices, but only concerning the power drive, and our illusions as to the motivations of our morality. It was a time critical of morality.

DR. EVANS: So Dr. Freud, in a sense, was taking another direction——

DR. JUNG: Yes; and then, moreover, sex being the main instinct, the predominating instinct in a more or less safe society. When the social conditions are more or less safe, then sexuality is apt to predominate because people are taken care of—they have their positions—they have enough food; no question of hunting and seeking food, or something like that—then it is quite probable that patients you meet have more or less all a certain sexual complex.

DR. EVANS: So this is, in a sense, the drive in that particular society most likely to be inhibited?

DR. JUNG: It is a sort of finesse, almost, when you find out that somebody has a power drive and their sex only serves the purpose of power. For instance, a charming man whom all women think is the real hero of all hearts; he is a power-devil, like a Don Juan, you know. The woman is not his problem; his problem is how to dominate. So, in the second place after sex comes the power drive. And even that is not the end.

Text pages 56–62

DR. EVANS: Now going fundamentally through the development of Dr. Freud's theory which, of course, was a significant factor as you say in the development of many of your own early ideas. Dr. Freud, of course, talked a great deal about the unconscious.

DR. JUNG: As soon as research comes to a question of the unconscious, then things become necessarily blurred, because the unconscious is something which is really unconscious; and so you have no object, you see nothing; you only can make inferences, you know, and so we have to create a model of this possible structure of the unconscious because we can't see it. Now he came to the concept of the unconscious chiefly on the basis of the same experience I have made in the association experiment; namely, that people reacted—they said things, they did things —without knowing that they did it or said it. This is something you can observe experimentally in the association experiment where people cannot remember afterward what they did or what they said in the moment where a stimulus word hits the complex. With the reproduction in the so-called reproduction experiment you go through the whole list, and you will see that the memory fails where there was a complex reaction. That is the simple fact Freud had based his idea of the unconscious on. Because that is what we can see, time and again, when people make a mistake in speech or they say something which they didn't mean to say; they just make ridiculous mistakes. There is no end of stories, you know, about how people can betray themselves by saying something they didn't mean to say at all, yet the unconscious meant them, meant them to say just that thing. For instance, when you want to express your sympathy at a funeral, you go to someone and you say "I congratulate you"—that's pretty painful, you know, that that happens, and it is true. This is something that goes parallel now with the whole school of the Salpetrière, in Paris. There was Pierre Janet, who has worked out that side of the unconscious reactions quite particularly. Now Freud refers very little to Pierre Janet, but I have studied with Pierre Janet in Paris, and he has formed my ideas very much. He was a first-class observer, though he had no dynamic theory about—no psychological theory. It was a sort of physiological theory of unconscious phenomena—the so-called *abaissement du niveau mentale*. There is a certain depotentiation of the tension of consciousness. A content sinks below the level of consciousness and thus becomes unconscious. That is Freud's view too, only he says it sinks down because it is helped—it is repressed from above. That was my first point of difference with Freud. I said there are cases in my observations where there was no repression from above. But the thing itself is true; those contents that became unconscious had withdrawn all by themselves; they were not repressed. On the contrary, they have a certain autonomy. There I discovered the concept of autonomy in that these contents that disappear have the power to move independently from my will. Either they appear when I want to say something definite, they interfere and say and speak themselves instead of what I wanted to say, or they make me do something which I didn't want to do at all, or they withdraw in the moment that I want to use them. They certainly disappear.

DR. EVANS: And this is independent of any of the, you might say in a sense, pressures on the consciousness as Freud——

DR. JUNG: Yes, there can be such cases, sure enough. But, beside them, there are also the cases that show that the unconscious contents acquire a certain independence. And all mental contents that have a certain feeling tone that is emotional, have the value of an emotional affect—have the tendency to become autonomous. So you see, anybody in an emotion will say and do things which he cannot vouch for. He must excuse himself that he was in a state; he was *non compos mentis*.

DR. EVANS: . . . Dr. Freud recognized that in a sense the individual is born entirely a victim of what he later called the id, a total sort of unconsciousness, undeveloped, sort of a total animal organism and I think it was not very easily seen where all these drives—all these instincts—came from——

DR. JUNG: Nobody knows where instincts come from. They are there—you find them. It is a story that was played out millions of years ago. There sexuality has been invented, and I wasn't there, so I don't know how this happened. Feeding has been invented very much longer ago than even sex; and how and why it has been invented I don't know. So we don't know where that thing comes from. It is quite ridiculous, you know, to speculate about such an impossibility. So the question is only—where do those cases come from where instinct does not function? That is something within our reach, because we can study the cases where instinct does not function.

DR. EVANS: Could you give us some rather specific examples of what you mean of cases where the instinct does not function?

DR. JUNG: Well, you see, instead of instinct, which is a habitual form of activity—take any other form of habitual activity. Once, suddenly, a thing doesn't function. Take a singer who is absolutely controlling his voice and suddenly he can't sing, or take another thing, a man who writes fluently, who suddenly makes a ridiculous mistake. There his habit doesn't function. You see, when you ask me something, I'm supposed to be able to react to you. Suddenly I am *bouche béante* [open-mouthed], words fail me. Or if you succeed to touch one of my complexes, you can see that I am absolutely perplexed; I am *depossedé*.

DR. EVANS: We haven't seen you very perplexed yet, Dr. Jung.

DR. JUNG: Or look at exam psychology, you know; a fellow who knows his stuff quite well; the professor asks him something and he cannot say a word.

DR. EVANS: Now going a little further—another part of Dr. Freud's theory, of course, which became very important, as we have already alluded to—was the idea of the conscious; that out of this sort of unconscious instinctual structure that the individual has, that out of this emerges as a result of this structure's—and of course the word "structure" has to be put in quotation marks, as you point out it is only a model—this structure comes into contact with reality in the world, it ultimately begins to emerge as the consciousness, then a man's conception almost of himself. Freud called this structure the ego. His ego, then, was really the product of reality—in a sense also perhaps the product of frustration as the individual developed.

DR. JUNG: That he has an ego at all, that is your question? Ah, that is again such a case—I haven't been there when it was invented, although you see you can observe it to a certain extent with a child. Because a child definitely begins in a state where there is no ego, then about the fourth year or before, but about that time, the child develops a sense of ego—"I, myself." That is in the first place a certain identity with the body. For instance, when you ask primitives, they emphasize always the body. For instance, you ask—"who has brought this thing here"—the Negro will say "Ibrought it," no accent on the "I," simply "brought it." You say—but have YOU brought it? and then he says, "in here, ME, ME, yes, I, MYSELF," this given object, this thing here. So the identity, you know, with the body is one of the first things which makes an ego; it is the spatial separateness that induces, apparently, the concept of an ego. Then, of course, there are lots of other things; later on it is mental differences, personal differences of all sorts, etc. You see, the ego is continuously building up; it is not a finished product, never; it builds up. You see, no year passes when you do not discover a new little aspect in which you are more ego than you thought.

DR. EVANS: Now as far as the influences in the process of developing of this ego on later personality, later over all organization of the individual as he tries to find himself in a fuller sense; of course there has been much discussion about how the early years influence the individual in these early formative years of the ego. For example, one of the most extreme views that we have—of one of the very men that you knew around Dr. Freud whom we mentioned a moment ago—Otto Rank. He spoke of the birth trauma; he said that the actual trauma of being born would leave a very powerful impact on the ego and show itself throughout the life of the person.

DR. JUNG: I should say that it is very important for an ego that it is born; this is highly traumatic, you know, when you fall out of heaven.

DR. EVANS: Do you take literally Otto Rank's position of the birth trauma as having, you might say, a psychological effect?

DR. JUNG: Of course, for instance, if you are a believer in Schopenhauer's philosophy, you say it is a hellish trauma to be born. Well, there is a Greek saying: "It is beautiful to die in youth, but the most beautiful of all things is not to be born." Philosophy, you see.

DR. EVANS: But you don't take this as a literal psychic event?

DR. JUNG: Don't you see, this is an event that happens to everybody that exists—that he once has been born. So everybody who is born has undergone that trauma, so the word

---

trauma has lost its meaning. It is a general fact, of which you cannot say—"It is a trauma"; it is just a fact, because you cannot observe a psychology that hasn't been born; only then you could say what the birth trauma is. Until then you cannot even speak of such a thing; it is just a lack of epistemology.

Text pages 49–54

DR. EVANS: Going a little further—in the more or less orthodox psychoanalytic view, as you well know, there is a great deal of attention paid to what Freud called psychosexual development in that, step by step, the individual encounters a series of problems and he must resolve these problems to keep going further and further, and of course one of the earliest problems the individual seems to have centers around, you might say, primitive satisfactions of oral sorts, or at the oral level—sort of, levels that seem to be——

DR. JUNG: In other words, that nutritional instinct was more important than sex. That's not very interesting.

DR. EVANS: Do you interpret this, then, as a sort of nutritional hunger drive? You do not look at this, then, as a sort of early representation of sexual drive?

DR. JUNG: I think, you see, when you say one of the first interests and foremost interests is to feed. That doesn't need such a peculiar kind of terminology like "oral zone." Of course they put it into the mouth——

DR. EVANS: Then you look at it in a much simpler sense.

DR. JUNG: Science consists to a great extent of mere talk.

DR. EVANS: Now Dr. Jung, we have been talking about the early oral level and as you pointed out, you wanted to look at it rather literally as a sort of hunger drive or drive for nutrition. Another rather fundamental point in development of the ego, as the more or less orthodox psychoanalytic view follows, is that this is followed by another critical level— an anal level of development where, in a sense, the whole process of the basic alimentary processes are more or less a basis of much frustration as the child is toilet-trained, etc. Now again, these later are interpreted as being very significant from the standpoint of having broad sexual implications. Would you regard this level a so-called "anal level" in this sense?

DR. JUNG: Well, one can use such terminology because it is a fact that children are exceedingly interested in all orifices of the body and in doing all sorts of disgusting things, and sometimes such a peculiarity keeps on into later life. It is quite astonishing what you can hear in this respect. Now it is equally true that people who have such preferences also develop a peculiar character. In early childhood a character is already there. You see a child is not born *tabula rasa* as one assumes. The child is born as a high complexity, with existing determinants that never waver through the whole life, and that give the child his character. Already, in earliest childhood, a mother recognizes the individuality of her child; and so if you observe carefully you see a tremendous difference, even in very small children. And these peculiarities express themselves in every way, so first the peculiarities express themselves in all childish activities—in the way it plays, in the things it is interested in. There are children who are tremendously interested in all moving things—in the movement chiefly; all things they see that affect the body. And so they are interested in what the eyes do, what the ears do, how far you can bore into the nose with your finger, you know. They will do the same with the anus; they will do whatever they please with their genitals. For instance, when I was in school we once stole the class book where all the punishments were noted, and there our professor of religion had noted: "So-and-so punished with two hours because he was toying with his genitals during the religious hour." These interests express themselves in a typically childish way in children. And later on they express themselves in other peculiarities which are still the same, but it doesn't come from the fact that they once had done such and such a thing in childhood. It is the character that is doing it. There is a definite inherited complexity, and if you want to know something about possible reasons you must go to the parents. So in any case of a child's neurosis, I go back to the parents and see what there is going on; because children have no psychology of their own, literally taken. They are so much in the mental atmosphere of the parents, so much *en participation mystique* with the parents, they are imbued by the paternal or maternal atmosphere, and they express these influences in their childish ways. So for instance, take an illegitimate child. They are particularly exposed to environmental difficulties, for instance, the misfortune of the mother, etc., etc., and all the complications. Such a child will miss, for instance, a father. Now in order to compensate for this it is just as if they were choosing or nominating a part of their body for a father, instead of the father, and they develop, for instance, masturbation. That is very often so with illegitimate children—they become terribly autoerotic, even criminal.

DR. EVANS: Now along the lines of the role of the parents in development; of course one of the central parts of the so-called psychosexual development in the more or less orthodox psychoanalytic theory is the so-called Oedipus complex. I believe the term "complex" was yours originally.

DR. JUNG: That is just what I call an archetype. That is the first archetype Freud has discovered; the first and the only one. He thinks that this *is* the archetype. Of course there are many such archetypes. You look at Greek mythology and you find them, any amount of them. Or look at dreams and you find any amount of them. But incest was so impressive to him that he even has chosen the term Oedipus complex, because that was one of the outstanding examples of an incest complex, and it is only in the masculine form, mind you, because women have an incest complex too, but there it was not an Oedipus, so it is something else. And so it is only a term for an archetypal way of behavior, in the case of a man's relation, say, to his mother. But it also means to his daughter because whatever he was to the mother, he will be it to the daughter too. It can be this way or that way. It depends.

DR. EVANS: Then you will accept, in other words, the Oedipus complex but not as being the only important such influence. You see this as just one of many.

DR. JUNG: One of the many, many ways of behavior. Oedipus gives you an excellent example of the behavior of an archetype. It is always a whole situation. There is a mother, there is a father, there is a son, there is a whole story of how such a situation develops and to what end it leads finally. That is an archetype. An archetype always is a sort of abbreviated drama. It begins in such and such a way, it extends to such and such a complication, and finds its solution in such and such a way. That is the usual form. As for instance take the instinct of building the nests of birds. In the way they build the nest— there is the beginning, the middle, and the end. It is built just to suffice for a certain number of young. The end is already anticipated. That is the reason why, in the archetype itself there is no time; it is a timeless condition where beginning, middle, and end are just the same; they are all given in one. That is only a hint to what the archetype can do, you know. But that is a complicated question.

DR. EVANS: Taking rather specifically the so-called Oedipus complex, a rather common statement, again by the fairly orthodox psychoanalytic view, was that in a sense these early family behavior patterns for the mother, the father, etc., more or less are relived over and over again. For example, the young man gets married. In a sense he may to some degree react to his wife as he did to his mother, or he may be searching for someone like his mother; or likewise the daughter will be searching for a father. This will be repeated over and over again. In a sense this is the heart of what you have been theorizing. Now this type of recapitulation—of this very early Oedipus situation——

DR. JUNG: Yes.

DR. EVANS: Does this fit in with your conceptions?

DR. JUNG: No. You see, Freud speaks of the incest complex just in the way you describe, but he omits completely the fact that with this Oedipus complex is always given the contrary—namely, the resistance against it. For instance, if the Oedipus were really predominant, we would have been suffocated in incest half a million years ago, at least. But there is a compensation. Namely, in all of the early levels of civilization you find the marriage laws, namely exogamic laws. The first form, the most elementary form, is that the man can only marry his cousin on the maternal side. The next form is that the man can only marry his cousin in the second degree, namely from the grandmother. There are four systems; quarternary systems, and systems of eight and of twelve, and in China there are still some traces of a twelve-system, also of a six-system. And those are developments beyond the incest complex and against the incest complex. Now if sexuality is predominant, particularly incestual sexuality, how can it develop? These things have developed in a time long before there was any idea of the child, say, of my sister. That's all wrong. On the contrary it was a royal prerogative, as late as the Achaemenite kings in Persia, and the Egyptian Pharaohs. If the Pharaoh has had a daughter from his sister, he married that daughter and again had a child with her and again married his granddaughter. Because that was royal prerogative. And so you see, the preservation of the royal blood is always a sort of attempt at the highly appreciated incestuous restriction of the numbers of ancestors, because this is loss of ancestors. Now you see that must be explained too. It is not only the one thing but is also its compensation. You know this plays a very great role in the history of human civilization. So Freud is always inclined to explain these things by external influences. For instance, you would not feel hampered in any way if there were not a law against it. No one is hampered by oneself. And that's what he never could admit to me.

Text page 62

DR. EVANS: Now of course, apropos that point, in his writing later of course, in addition to the ego; he introduces a sort of part ego which he called the "superego." Broadly speaking, of course——

DR. JUNG: Yes, that is the superego, namely that codex of what you can do and what you cannot do.

DR. EVANS: Built-in prohibitions——

DR. JUNG: Yes, but Freud doesn't see that it is *in* himself, he has it in himself. Otherwise there could be no balance in the individual. And who in hell would have invented the Decalogue? That is not invented by Moses, but that is the eternal truth in man himself, because he checks himself.

Text pages 65–67

DR. EVANS: Now you mentioned that among the archetypes that are to you very important of course, would be one something similar in a broad sense to the Oedipus situation. Now to perhaps clarify this for many of us who perhaps would like to be a little bit more specific about the concept archetype, perhaps you might, if you would, elaborate a little bit on the concept archetype. Exactly what it means. Of course you have alluded very clearly——

DR. JUNG: Well, you know what a behavior pattern is. The way in which say a weaverbird builds its nest. That is an inherited form in him which he will apply. Or certain symbiotic phenomena, between insects and plants. They are inherited patterns of behavior. And so man has, of course, an inherited scheme of functioning. You see, his liver, his heart, all his organs, and his brain will always function in a certain way, following its pattern. You and I have a great difficulty of seeing it because we cannot compare. There are no other similar beings like man that are articulate that could give an account of their functioning. If that were the case we could—I don't know what —but because we have no means of comparison we are necessarily unconscious about our own conditions; but it is quite certain that man is born with a certain functioning; a certain way of functioning; a certain pattern of behavior, and that is expressed in the form of archetypal images, or archetypal forms. For instance, the way in which a man should behave is given by an archetype. Therefore you see the primitives tell such stories. A great deal of education goes through storytelling. For instance, they call in a palaver of the young men and two older men perform before the eyes of the younger all the things they should not do. Then they say, "Now that's exactly the thing you shall not do." Another way is, they tell them of all the things they should not do, like the Decalogue—"Thou shalt not." And that is always supported by mythological tales. For instance, our ancestors have done so-and-so, and so you shall do. Or—such and such a hero has done so-and-so, and that is your model. For instance, teaching of the Catholic Church—there are several thousand saints. They serve as models; they have their legends; and that is Christian mythology. In Greece, you know, there was Theseus, there was Heracles—models of fine men, you know, and they teach us how to behave; they are archetypes of behavior.

Text pages 70–74

DR. EVANS: In other words, in a sense——

DR. JUNG: Or, for instance, you see—taking a more concise archetype like the archetype of the ford; the ford through a river. Now that is a whole situation. You have to cross a ford, you are in the water, there is an ambush, or there is a water animal, say a crocodile or something like that—and there is a danger, and something is going to happen, and the question is, how you escape. Now this is a whole situation that makes an archetype. And that archetype has now a suggestive effect upon you. For instance, you get into a situation, you don't know what the situation is, you suddenly are seized by an emotion or by a spell, and you behave in a certain way you have not foreseen at all; you do something quite strange to yourself.

DR. EVANS: We'll call it spontaneous.

DR. JUNG: Quite spontaneous; and that is done through that archetype that is concerned. Of course we have a famous case in our Swiss history of the King Albrecht, who was murdered in the ford of the Reuss not very far from Zurich. His murderers were hiding behind him for the whole stretch from Zurich to the Reuss—quite a long stretch—and deliberated and couldn't come together whether they wanted to kill the king or not. The moment the king rode into the ford, Johannes Parricida, the father murderer, shouted, "Why do we let him abuse us?" —and used a swear word. And they killed him, because this was the moment they were seized, this was the *right* moment. So you see, when you have lived in primitive circumstances, in the primeval forest, among primitive populations, then you know that phenomenon. You are seized with a certain spell, then you do a thing that is unexpected. Several times when I was in Africa I went into such situations where I was amazed afterwards. One day I was in the Sudan, and it was really a very dangerous situation—had been a dangerous situation which I didn't recognize at the moment at all, but I was seized with a spell and I had done something which I wouldn't have expected and I couldn't have invented it. You see the archetype is a force; it has an autonomy; it can suddenly seize you—it is like a seizure. So, for instance, falling in love at first sight, that is such a case. You see, you have a certain image in yourself without knowing it, of the woman—of *the* woman. Now you see that girl, or at least a good imitation of your type, and instantly you get the seizure and you are gone. And afterward you may discover that it was a hell of a mistake. So you see a man is quite capable, or is intelligent enough to see that the woman of his choice was no choice—he has been captured; he sees that she is no good at all, that she is a hell of a business, and he tells me so, and he says, "for God's sake, doctor, help me to get rid of that woman"—and he can't, he is like clay in her fingers. That is the archetype; that is the archetype of the anima, and he thinks it is all his soul, you know; like the girls, you know, when a man sings very high, she thinks he must have a very wonderful spiritual character because he can sing high C, and she is badly disappointed when she marries that particular "letter." Well, that's the archetype of the animus.

DR. EVANS: Now Dr. Jung, to be even a little bit more specific; of course you have suggested that in our society, in all societies, there are symbols that in a sense direct or determine what a man does. Then of course you also suggest that somehow these symbols become inborn—become part, inbred.

DR. JUNG: They don't become; they *are*. They are to begin with. You see, we are born into a pattern; we *are* a pattern. We are a structure that is preestablished through the genes.

DR. EVANS: Would you say, then, that this is just a higher order of an instinctual pattern, such as, for instance, say in a bird building a nest, and that is how you look at it?

DR. JUNG: It is a biological order of our mental functioning, as for instance our biological-physiological function follows a pattern. The behavior of any bird or insect follows a pattern, and that is the same with us. Man has a certain pattern that makes him specifically human, and no man is born without it. We are only deeply unconscious of these facts because we live by all our senses and outside of ourselves. If a man could look into himself he could discover it. When a man discovers it in our days he thinks he is crazy—and he may be crazy.

DR. EVANS: Now would you say the number of such archetypes are limited, pre-fixed, or can the number increase?

DR. JUNG: Well, I don't know what I do know about it; it is so blurred. You see, we have no means of comparison. We know, you see, there is a behavior, say like incest; or there is a behavior of violence—certain kind of violence; a behavior of panic; a behavior of power, etc. Those are areas, as it were, in which there are many variations. It can be expressed in this way or that way, you know. And they overlap, and often you cannot say where the one form begins or ends. It is nothing

concise, because the archetype in itself is completely unconscious and you only can see the effects of it. You can see, for
instance, when you know a person possessed by an archetype,
then you can divine, and even prognosticate possible developments. Because when you see that the man is caught by a
certain type of woman in a certain very specific way, you know;
he is caught by the anima. Then the whole thing will have such
and such complications, such and such developments, because
it is typical. The way the anima is described is exceedingly
typical. The interesting thing is, for instance, I don't know if
you know Rider Haggard's *She*, or *L'Atlantide* by Benoit. Those
are anima types, and they are quite unmistakable—*c'est la
femme fatale*.

DR. EVANS: Now just to be a little bit more specific, Dr.
Jung, you have used the concept "anima" and "animus" and
you are now identifying in terms of sex—male or female. Now
I wonder if you could elaborate perhaps even more specifically
on these terms? Take the term "anima" first of all. Now this
again is part of the inherited nature of the individual.

DR. JUNG: Well, this is a bit complicated, you know. The
anima is an archetypal form, expressing the fact that a man has
a minority of feminine or female genes. That is something that
doesn't appear or disappear in him; that is constantly present,
and it works as a female in a man; therefore already in the
sixteenth century the Humanists have discovered that man has
an "anima" and that each man carries his female with himself,
they said. So that is not a modern invention. The same is the
case with the animus; that is a masculine image in a woman's
mind which is sometimes quite conscious—sometimes it is not
conscious, but it is called into life the moment that woman
meets a man who says the right things, and then because he
says it, it is all true, and he is the fellow, no matter what he is.
Those are particularly well founded archetypes, those two. And
there you can lay hands on the basis, as it were, of the archetype. They are extremely well defined.

Second
Interview

Text pages 79–84

DR. EVANS: Dr. Jung, we have been discussing in some detail some of the factors in the development of the personality of the individual, and you have very kindly elaborated for us some of your fundamental concepts such as the archetype, at least simply what it means—certain types of archetypes, such as the anima and animus, and we tried to show, perhaps, in our discussion some of the ways in which your ideas may have differed from those of Dr. Freud. Now another concept or idea that seems to be a very interesting one in your work, at least as I see it, is the term or concept "persona." This seems to have a lot of relevance to the daily living of the individual. I wonder if you would mind telling us a little bit about how you construe this term "persona."

DR. JUNG: Well, this is a practical concept we need in elucidating people's relations. I noticed with my patients, par-

185

ticularly with people that are in public life, that they have a certain way of presenting themselves. For instance, take the doctor. He has a certain way—for instance, he has good bedside manners, and he behaves as one expects a doctor to behave. He may even identify himself with it and believe that he is what he appears to be. He must appear in a certain form, or else people won't believe that he is a doctor. And so when he is a professor, he is also supposed to behave in a certain way so that it is plausible that he is a professor. So the persona is partially the result of the demands society has. On the other side, it is a compromise with what one likes to be, or as one likes to appear. Take, for instance, a parson. He also has his particular manner and of course runs into the general expectation, and he behaves also in another way, combined with his persona that is forced upon him by society, in such a way that also his fiction of himself, his idea about himself, is more or less portrayed or represented. So the persona is a certain complicated system of behavior which is partially dictated by society and partially dictated by the expectations or the wishes one nurses oneself. Now this is not the real personality; in spite of the fact that people will assure you that it is all quite real and quite honest, yet it is not. Such a performance of the persona is quite all right, as long as you know that you are not identical with the way in which you appear. But if you are unconscious of this fact, then you get into, sometimes, very disagreeable conflicts; namely, people can't help noticing that at home you are quite different from what you appear to be in public. People who don't know it stumble over it in the end. They deny that they are like that, but they are like that—they are it. Then you don't know, now which is the real man? Is he the man as he is at home or in intimate relations, or is he the man that appears in public? It is a question of Jekyll and Hyde. Occasionally there is such a difference that you would almost be able to speak of a double personality. And the more that is pronounced, the more people are neurotic. They get neurotic because they have two different ways; they contradict themselves all the time, and inasmuch as they are unconscious of themselves, they don't know it. They think they're all one; everybody sees that they are two. Some know him only from one side; others know him only from the other side. And then there are situations that clash, because the way you are creates certain situations with people in your relations, and these two situations don't chime, in fact—they are just dishonest. And the more that is the case, the more the people are neurotic.

DR. EVANS: Actually, would you say that the individual

may even have more than two personas? In other words, could
he possibly——

DR. JUNG: We can't afford very well to play more than two
roles, but there are. There are cases where people have up to
five different personalities, in cases of dissociation of personality, where, for instance, the one person—call him person A—
doesn't know of the existence of the person B, but B knows of
A. There may be a third personality, C, that doesn't know of the
two others. There are such cases in the literature, but they
are rare.

DR. EVANS: Very rare.

DR. JUNG: In ordinary cases, it's just an ordinary dissociation of a personality. One calls that a systematic dissociation, in
contradistinction to the chaotic or unsystematic dissociation you
find in schizophrenia.

DR. EVANS: Do you distinguish between the term "persona"
and the term "ego"? In other words, as you see them, they are
two different things?

DR. JUNG: Yes.

DR. EVANS: What is the difference between the term "ego"
as you see it and the term "persona"?

DR. JUNG: Well, you see, the ego is supposed to be the
representative of the real person. For instance, in the case
where B knows of A, but A doesn't know of B, in that case one
would say the ego is more on the side of B, because the ego has
a more complete knowledge, and A is a split-off personality.

DR. EVANS: Another term that you use in this respect is the
term "self."

DR. JUNG: That I do. We all speak of "I," but it is not the
same "I"; it is only in part the same "I." The body is the same,
but the mind is not the same, or the character is not the same.

DR. EVANS: You also use the term "self." Now the word
"self"—does it have a different meaning than "ego" or "persona"?

DR. JUNG: Yes. When I say "self," then you mustn't think
of "I, myself," because that is only your empirical self, and that
is covered by the term "ego," but when it is a matter of "self,"
then it is a matter of a personality that is more complete than
the ego, because the ego only consists of what you are conscious
of, what you know to be yourself. For instance, in our example,
B that knows A but A doesn't know B, B is relatively in the
position of the self; namely, the self is, on the one side, the ego,
on the other side, the unconscious personality which is in the
possession of everybody—not in the possession. Very often it is
just the other way around; the unconscious is in the possession

of consciousness. That is a different case. Now you see, while I am talking, I am conscious of what I say; I am conscious of myself, yet only to a certain extent. Quite a lot of things happen. For instance, I make gestures, I'm not conscious of them. They happen unconsciously. You can see them. I may say or use words and can't remember at all having used those words, or even at the moment I am not conscious of them. So any amount of unconscious things occur in my conscious condition. I'm never wholly conscious of myself. While I am trying, for instance, to elaborate an argument, at the same time there are unconscious processes that continue, perhaps a dream which I had had last night, or a part of myself thinks of God knows what, of a trip I'm going to take, or of such and such people I have seen. Or say, when I am writing a paper, I am continuing writing that paper in my mind without knowing it. You can discover these things, say, in dreams, or if you are clever, in the immediate observation of an individual. Then you see in the gestures, or in the expression in the face, that there is what one calls *une arrière pensée*, something behind consciousness. You have finally the feeling, well, that man has something up his sleeve, and you can even ask him, "What are you really thinking of? You are thinking all the time something else." Yet he is not conscious of it. Or, he may be. There are, of course, great individual differences. There are individuals who have an amazing knowledge of themselves, of the things that go on in themselves. But even those people wouldn't be capable of knowing what is going on in their unconscious. For instance, they are not conscious of the fact that while they live a conscious life, all the time a myth is played in the unconscious, a myth that extends over centuries, namely, archetypal ideas— a stream of archetypal ideas that goes on through one individual through the centuries. Really it is like a continuous stream and that comes into the daylight in the great movements, say in political movements or in spiritual movements. For instance, in the time before the Reformation, people dreamt of the great change. That is the reason why such great transformations could be predicted. If somebody has been clever enough to see what is going on in people's mind, in the unconscious mind, he would be able to predict it. For instance, I have predicted the Nazi rising in Germany through the observation of my German patients. They had dreams in which the whole thing was anticipated, and with considerable detail. And I was already absolutely certain, in the years before Hitler—before Hitler came in the beginning—I could say the year, in the year 1919, I was sure that something was threatening in Germany, something very big, very catastrophic. I only knew it through the observation of the unconscious. There is something very particular in the different nations. It is a peculiar fact that the archetype of the anima plays a very great role in Western literature, French and Anglo-Saxon. Not in Germany; there are exceedingly few examples in German literature where the anima plays a role. You know, that simply comes of the fact that not one woman is buried unless she is buried as *alt Kaminfegersgattin,* at least. She must have a title; otherwise she hasn't existed. And so it is just as if—now mind you, this is a bit drastic, but it illustrates my point—in Germany there really are no women; there is Frau Doktor, Frau Professor, Frau the grandmother, the mother-in-law—the grandfather, father, the son, the daughter, the sister. No—*la femme n'existe pas.* That is the idea, you see. Now that is an enormous, enormously important fact which shows that in the German mind there is going on the particular myth, something very particular, and psychologists really should look out for these things. But they prefer to think that "I am a prophet." Ha!

DR. EVANS: This is of course a very interesting and remarkable set of statements here. How would you look at Hitler in this light? Would you see him as a personification, as symbol of the father?

DR. JUNG: No, not at all. I couldn't possibly explain that very complicated fact that Hitler represents. It is too complicated. You know, he was a hero figure, and a hero figure is far more important than any fathers that have ever existed.

DR. EVANS: Much broader——

DR. JUNG: Not at all; he was a hero in the German myth. And mind you, a religious hero. He was a savior; he was meant to be a savior. That is why they put his photo upon the altars even. Or somebody declared on his tombstone that he is happy that his eyes had beheld Hitler, and now he can lie in peace. He was just a hero myth.

DR. EVANS: Now, getting back to the idea of the self, the self incorporates these unconscious factors.

DR. JUNG: The self is merely a term that designates the whole personality. The whole personality of man is indescribable. His consciousness can be described; his unconscious cannot be described because the unconscious, and I repeat myself, is always unconscious. It is really unconscious; he really does not know it. And so we don't know our unconscious personality. We have hints, we have certain ideas, but we don't know it really. Nobody can say where man ends. That is the beauty of it, you know; it's very interesting. The unconscious of man can reach God knows where. There we are going to make discoveries.

Text pages 91–103

DR. EVANS: Now another set of ideas, which, of course, are very, very well known to the world, but of course, you have originated, center around the terms "introversion" and "extroversion." I know that you are aware of the fact that these terms have now become so widely known that the man in the street is using these terms, constantly, in describing his wife or his friend, etc.; it has become probably the most used psychological concept by the layman that we have.

DR. JUNG: Like the word "complex." I have invented it, too, you know, from the association experiments. This is simply practical, because there are certain people who definitely are more influenced by their surroundings than by their own intentions, while other people are more influenced by the subjective factor. Now you see the subjective factor, that's very characteristic, was understood by Freud as a sort of pathological auto-eroticism. Now this is a mistake. The psyche has two conditions, two important conditions. The one is the environmental influence and the other is the given fact of the psyche as it is born. The psyche is by no means *tabula rasa* but a definite mixture and combination of genes, and they are there from the very first moment of our life, and they give a definite character, even to the little child; and that is a subjective factor, looked at from the outside. Now if you look at it from the inside, then it is just so as if you would observe the world. When you observe the world you see people, you see houses, you see the sky, you see tangible objects; but when you observe yourself within, you see moving images—a world of images, generally known as fantasies. Yet these fantasies are facts. You see, it is a fact that the man has such and such a fantasy, and it is such a tangible fact, for instance, that when a man has a certain fantasy, another man may lose his life, or a bridge is built—these houses were all fantasies. Everything you do here, all of the houses, everything, was fantasy to begin with, and fantasy has a proper reality. That is not to be forgotten; fantasy is not nothing. It is, of course, not a tangible object, but it is a fact, nevertheless. It is, you see, a form of energy, despite the fact that we can't measure it. It is a manifestation of something, and that is a reality. That is just a reality. As for instance, the peace treaty of Versailles, or something like that. It is no more—you can't show it, but it has been a fact. And so psychical events are facts, are realities; and when you observe the stream of images within, you observe an aspect of the world, of the world within. Because the psyche, you know, if you understand it, as a phenomenon that takes place in so-called living bodies, then it is a quality of matter, as our body consists of matter. We discover that this matter has another aspect, namely, a psychic aspect. And so it is simply the world from within, seen from within. It is just as though you were seeing into another aspect of matter.

That is an idea that is not my invention. The old credos already talked of the *spiritus insertus atomis,* namely, the spirit that is inserted in atoms. That means psychic is a quality that appears in matter; it doesn't matter whether we understand it or not. But this is the conclusion we come to, if we draw conclusions without prejudices. And so you see, the man who is going by the external world, by the influence of the external world—say, society or sense perceptions—thinks that he is more valid, you know, because this is valid, this is real, and the man who goes by the subjective factor is not valid, because the subjective factor is nothing. No, that man is just as well based, because he bases himself upon the world from within. And so he is quite right even if he says, "Oh, it is nothing but my fantasy." And of course, that is the introvert, and the introvert is always afraid of the external world; he will tell you when you ask him. He will be apologetic about it; he will say, "Yes, I know, those are my fantasies." And he has always resentment against the world in general. Particularly America is extroverted like hell. The introvert has no place, because he doesn't know that he beholds the world from within. And that gives him dignity, and that gives him certainty, because it is—nowadays particularly, the world hangs on a thin thread, and that is the psyche of man. Assume that certain fellows in Moscow lose their nerve or their common sense for a bit; and the whole world is in fire and flames. Nowadays we are not threatened by elemental catastrophies. There is no such thing in nature as an H-bomb; that is all man's doing. We are the great danger. The psyche is the great danger. What if something goes wrong with the psyche? And so it is demonstrated to us in our days what the power of psyche is, how important it is to know something about it. But we know nothing about it. Nobody would give credit to the idea that the psychical processes of the ordinary man have any importance whatever. One thinks, Oh, he has just what he has in his head. He is all from his surroundings. He is taught such and such a thing, believes such and such a thing, and particularly if he is well housed and well fed, then he has no ideas at all. And that's the great mistake, because he is just that as which he is born, and he is not born as *tabula rasa* but as a reality.

DR. EVANS: Of course, one of the very common, I think, misconceptions of your work among some of the writers in

America is that they have sort of characterized your discussion of introversion and extroversion as suggesting that the world is made up of only two kinds of people, introverts and extroverts. I'm sure you have been aware of this . . . and would you like to comment about that? In other words, would you perceive of the world being made up of only people who are extreme introverts and people who are extreme extroverts?

DR. JUNG. Bismarck once said, "God may protect me against my friends; with my enemies I can deal myself alone." You know how people are. They have a catchword, and then everything is schematized along that word. There is no such thing as a pure extrovert or a pure introvert. Such a man would be in the lunatic asylum. Those are only terms to designate a certain penchant, a certain tendency. For instance, the tendency to be more influenced by environmental influences, or more influenced by the subjective fact—that's all. There are people who are fairly well balanced and are just as much influenced from within as from without, or just as little. And so with all the finest classifications, you know, they are only a sort of *points de repère,* points for orientation. There is no such thing as a schematic classification. Often you have great trouble even to make out to what type a man belongs, either because he is very well balanced or he is very neurotic. Because when you are neurotic, then you have always a certain dissociation of personality. And then the people themselves don't know when they react consciously or when they react unconsciously. So you can talk to somebody, and you think he is conscious. He knows what he says, and to your amazement you discover after a while that he is quite unconscious of it, doesn't know it. It is a long and painstaking procedure to find out of what a man is conscious and of what he is not conscious, because the unconscious plays in him all the time. Certain things are conscious; certain things are unconscious, but you couldn't tell. You have to ask people now, "Are you conscious of what you say?" Or, "Did you notice?" And you discover suddenly that there are quite a number of things that he didn't know at all. For instance, certain people have many reasons; everybody can see them. They themselves don't know it at all.

DR. EVANS: Certainly, then, this whole matter of extremes —introvert and extrovert—as you say, it is a sort of a scheme or schematic approach to sort of hang an idea on. An approach, but as you say, it would be ridiculous to say . . .

DR. JUNG: My whole scheme of typology is merely a sort of orientation. There is such a factor as introversion; there is such a factor as extroversion. The classification of individuals means nothing at all. It is only the instrumentarium for what I call practical psychology, to explain, for instance, the husband to a wife, or vice versa. For instance, it is very often the case— or I might say it is almost a rule, but I don't want to make too many rules in order not to be schematic—that an introvert marries an extrovert for compensation, or another type marries a counter-type to complement himself. For instance, a man who has made a certain amount of money is a good business man, but he has no education. His dream is, of course, a grand piano at home, and artists, you know, painters or singers or God knows what, and intellectual people, and he marries accordingly a wife of that type, in order to have that, too. Of course, he has nothing of it. She has it, and she marries him because he has a lot of money. These compensations go on all the time. When you study marriages, you can see it easily. And, of course, we analysts have to deal a lot with marriages, particularly those that go wrong, because the types are too different sometimes, and they don't understand each other at all. But you see, the main values of the extrovert are anathema to the introvert, and he says, "To hell with the world, I think." His wife interprets this as his megalomania. While it is just so as if an extrovert would say to an introvert, "Now, look here, fellow; these here are the facts, this is reality, and he's right." And the other says, "But *I* think, *I* hold . . ."—and that sounds like nonsense to the extrovert because he doesn't know that the other one, without knowing it, is beholding an inner world, an inner reality, and he may be right, as he might be wrong, even if he found himself upon God knows what solid facts. Take, for instance, the interpretation of statistics. You can prove almost anything with statistics. What is more a fact than a statistic?

DR. EVANS: Now of course, with respect to your typology, you talk about function, thinking, feeling, intuition, sensation. You have again used these not as types at all, but as patterns that may tie in with these introverted and extroverted tendencies.

Dr. Jung, of course, tied in with your typology of introversion-extroversion, we of course know of your concepts of thinking, feeling, sensation, intuition; and of course it would be very interesting to hear some expansion of the meaning of these particular terms as related to the introvert-extrovert dichotomy.

DR. JUNG: Well, there is a quite simple explanation of these terms, and it shows at the same time how I arrived at such a typology; namely, sensation tells you that there is something. Thinking, roughly speaking, tells you what it is. Feeling tells you whether it is agreeable or not, to be accepted or not, accepted or rejected. And intuition—there is a difficulty. You

don't know, ordinarily, how intuition works. So, when a man
has a hunch, you can't tell exactly how he got that hunch, or
where that hunch comes from. It is something funny about
intuition. I will tell you a little story. I had two patients; the
man was a sensation type, the woman was an intuitive type.
Of course, they felt attraction. And so they took a little boat
and went out to the lake of Zurich. And there were those birds
that dive after fish, you know, and then after a certain time they
come up again, and you can't tell where they come up. And so
they began to bet who was the first to see the bird. Now you
would think that the one who observes reality very carefully—
the sensation type—would of course win out. Not at all. The
woman won the bet completely. She was beating him on all
points because by intuition she knew it before. How is that
possible? You know, you can really find out how it works by
finding the intermediate links. It is a perception, by intermediate links, and you only get the result of that whole chain
of associations. Sometimes you succeed in finding out but more
often you don't. So, my definition is—intuition is a perception,
by ways or means of the unconscious. That is near as I can
get. This is a very important function, because when you live
under primitive conditions, a lot of unpredictable things are
likely to happen. There you need your intuition because you
cannot possibly tell by your sense perceptions what is going to
happen. For instance, you are traveling in primeval forests. You
can only see a few steps ahead; you go by the compass, perhaps,
and you don't know what there is ahead. It is uncharted country. If you use your intuition, then you have hunches, and when
you live under such primitive conditions, you instantly are
aware of hunches. There are places that are favorable; there are
places that are not favorable. You can't tell for your life what
it is, but you better follow these hunches, because anything can
happen—quite unforeseen things. For instance, at the end of a
long day you approach a river. You don't know that there is a
river, yet when you come to the river, that is quite unexpected.
For miles there is no human habitation. You cannot swim
across; it is all full of crocodiles. So what? Such an obstacle
hasn't been foreseen, but it may be that you have had a hunch
that you remain in the least likely spot and that you wait for the
following day that you can build a raft or something of the
sort, or look out for possibilities. For instance, you also can
have intuitions—that constantly happens—in our jungle called
a city. You can have a hunch, something that's going wrong,
particularly when you are driving an automobile. For instance,
it is a day where nurses appear in the street. And they always

try to get something interesting, like a suicide, you know—to be run over, that's more marvelous, apparently. And then, you know, you get a peculiar feeling, really, at the next corner a second nurse runs in front of the automobile. A multiplicity of cases, that is a rule, you know, that such chance happenings come in groups. And so you see, we have constantly warnings or hints, that consist partly in a slight feeling of uneasiness, uncertainty, fear. Now under primitive circumstances you would pay attention to these things; they mean something. With us in our man-made, absolutely, apparently, safe conditions, we don't need that function so very much; yet we have seen it and used it. So you will find that the intuitive types—for instance, amongst bankers, Wall Street men, they follow hunches, you know, as do gamblers of all descriptions. You find the type very frequently among doctors because it helps them in their prognoses. Sometimes a case can look quite normal, as it were, and you don't foresee any complications, yet an inner voice says, "Now you look out here, there is something not quite all right." You cannot tell why or how, but we have a lot of subliminal perceptions—sense perceptions—and from those we probably draw a good deal of our intuitions. But that is perception by the way of the unconscious, and you can observe that with intuitive types. You see, intuitive types very often do not perceive by their eyes or by their ears; they perceive by intuition. For instance, once it happened that I had a woman patient in the morning, at nine o'clock. You see, I often smoke my pipe and have a certain smell of tobacco in the room or of cigar. And so she came and said, "But you begin earlier than nine o'clock; you must have seen somebody at eight o'clock." And I said, "How do you know?" But there had been a man there at eight o'clock already. And she said, "Oh, I just had a hunch that there must have been a gentleman with you this morning." I said, "How do you know it was a gentleman?" And she said, "Oh, well, I just had the impression; the atmosphere was just like a gentleman here." All the time, you know, the ashtray was under her nose, and there was a half-smoked cigar but she wouldn't notice it. So you see, the intuitive is a type that doesn't see, doesn't see the stumbling block before his feet, but he smells a rat for ten miles.

Text pages 104–109

DR. EVANS: Do you make a distinction between an intuitive extrovert and an intuitive introvert?

DR. JUNG: Yes, all those types cannot be alike.

DR. EVANS: More specifically, what would be an example of difference between an intuitive extrovert and an intuitive introvert?

DR. JUNG: Well, you know, you have chosen a somewhat difficult case, because one of the most difficult types is the intuitive extrovert. The intuitive extrovert you find in all kinds of bankers, gamblers; that is quite understandable. But the introvert variety is more difficult because he has intuitions as to the subjective factor—namely the inner world—and, of course, that is now very difficult to understand because what he sees are most uncommon things, and he doesn't like to talk of them if he is not a fool, because he would spoil his own game by telling what he sees. Because people won't understand it. For instance, once I had a patient, a young woman about twenty-seven or twenty-eight. Her first words were when I had seated her, she said, "You know, doctor, I come to you because I've a snake in my abdomen." What! "Yes, a black snake coiled up right in the bottom of my abdomen." And I must have made a rather bewildered face at her. She says, "You know, I don't mean it literally. But I should say it was a snake, it was a snake." Our further conversation a little later was that she said —that was about in the middle of her treatment that only lasted for ten consultations—she had foretold me, "I come ten times, then it's all right." I said, "How do you know?" "Oh, I've got a hunch," she said. At about the fifth or sixth hour she said, "Doctor, I must tell you, the snake has risen; it is now about here." A hunch. Then on the tenth day I said, "Now this is our last hour, and do you feel cured?" And she said, beaming, "You know, this morning it came up and came out of my mouth and the head was golden." Those were her last words. That same girl, when it comes to reality, she came to me because she couldn't hear the step of her feet anymore, because she walked on air, literally. She couldn't hear it, and that frightened her. And when she came to me I asked her for her address, and she said, "Oh, pension so-and-so. Well, it is not just called a pension, but it is a sort of pension." I had never heard of it. "I have never heard of that place," I said. And she said, "It is a very nice place. Curiously enough, there are only young girls there; very nice and very lively young girls, and they have a merry time. I often wish they would invite me to their merry evenings." And I said, "Do they amuse themselves all alone?" And she said, "No, there are plenty of young gentlemen coming in; they have a beautiful time, but they never invite me." It turned out that it was a private brothel. She was a perfectly decent girl, of a very good family, not from here. She had found that place, I don't know how, and she was completely unaware that they were all prostitutes. And I say, "For heaven's sake, you fell into a very tough place; you hasten to get out of it." That was her sensation; she didn't see reality, but she had hunches like anything. Such a person cannot possibly speak of her experiences because everybody would think she is absolutely crazy. I myself was quite shocked, and I thought, for heaven's sake, is that case a schizophrenia? You don't hear that kind of speech, but she assumed that the old man, of course, knows everything, and he even does understand such kind of language. So you see, when the introverted intuitive would speak what he really perceives, practically no one would understand him; he would be misunderstood. And so they learn to keep things to themselves, and you hardly ever hear them talking of these things. That is a great disadvantage, but it is an enormous advantage in another way, not to speak of the experiences they gather in that respect, but also in human relations. For instance, they come into the presence of somebody they don't know, and suddenly they have inner images, and these inner images give them a more or less complete information about the psychology of the partner. The case can also happen that they come into the presence of somebody that they don't know at all, not from Adam, and they know an important piece out of the biography of that person, and are not aware of it, and they tell the story, and then the fat is in the fire. So the introverted intuitive has in a way a very difficult life, although one of the most interesting lives, but it is difficult often to get into their confidence.

DR. EVANS: Yes, because you say they are afraid people will think they are sick.

DR. JUNG: The things that are interesting to them, are vital to them, are utterly strange to the ordinary individual, and a psychologist should know of such things. When people make a psychology, as a psychologist ought to do, the very first question—is he extroverted or is he introverted? He will look at entirely different things. Is he sensation type, is he intuitive type, is he thinking, is he feeling, because, you see, these things are complicated. They are still more complicated because the introverted thinker, for instance, is compensated by extroverted feeling, by inferior, extroverted, archaic, extroverted feeling. So an introverted thinker may be very crude in his feeling, like for instance the introverted philosopher that is always carefully avoiding women will be married by his cook in the end.

DR. EVANS: Now we can take your introvert-extrovert category in a sense and go through and describe the sensation-introvert type, sensation-extrovert type, thinking-introvert type, etc. In each case it stands for not a real category but simply, as you say, an approach—something to sort of help us study, a model as it were.

DR. JUNG: It is just a sort of skeleton to which you have to add the flesh—or say it is a country mapped out, you know, by triangulation points, and that doesn't mean the country consists of triangulation points; that is only in order to have an idea of the distances. And so it is a means to an end. It only makes sense, such a scheme, when you deal with practical cases. For instance, if you have to explain to an extrovert wife an introverted intuitive husband, that is a most painstaking affair because, you see, an extrovert-sensation type is furthest away from the inner experience or rational functions. She adapts and behaves according to the facts as they are, and she is always caught by those facts. She, herself, is those facts. But if the introvert is intuitive, to him that is hell, because as soon as he is in a definite situation, he tries to find a hole where he can get out. Because every given situation is just the worst that can happen to him. He is pinched and feels he is caught, suffocating, chained. He must break those fetters because he is the man that will discover a new field. He will plant that field, and as soon as the young plants are coming up, he's over, he's done, he's no more interested; he is all right, and others will reap what he has sown. When those two marry each other, there is trouble, I can assure you.

Text pages 139–143

DR. EVANS: Yes indeed. Now speaking—going back, incidentally, to the intuitive type. Of course, you are familiar with the work of Dr. J. B. Rhine at Duke University. Some of his work in extrasensory perception and clairvoyance, as he calls it—mental telepathy—some of the descriptions in his work sound quite a bit at times like the intuitive . . . and in terms of your analogy of this work, would you say that a person who has clairvoyance, would then be an intuitive type? For instance, in his experiment with telepathy . . .

DR. JUNG: That's quite probable. Or it can be a sensation type—say an extrovert-sensation type that is very much influenced by the unconscious. He has introverted intuition in his unconscious. There are two groups, the rational group and the irrational group. The rational group is thinking and feeling. The ideal thinking is a rational result. Feeling is also a rational result, rational values. That is differentiated feeling, whereas sensation must need be irrational because it may not prejudice facts; it shall not prejudice facts. The real ideal perception is that you have an accurate perception of the things as they are without additions or corrections. On the other side, the intuition doesn't look at the things as they are. That is prison; that is anathema to the intuition. It looks ever so shortly at things as they are and makes off into an unconscious process at the end of which he has seen something nobody else will have seen. Now so it is these people who yield the best results are always those people who are introverted, or where introverted intuition comes in. But that is a side aspect of it; it is not interesting. There is another question far more interesting, namely, the terms they use. Rhine himself uses them—recognition, telepathy, etc. They mean nothing at all. They are words, but he thinks he has said something when he says telepathy.

DR. EVANS: The word itself is not a description of the process.

DR. JUNG: Not a description. It means nothing, nothing at all.

DR. EVANS: Now, of course, a lot of the things that you are describing, I think often scientists will say, "This is due to chance." Chance occurrences, chance factors. Rhine in his own work used statistical methods. We find this happening more often than would be expected by chance.

DR. JUNG: Well, you see, he proves that it is more than chance; it is statistically plausible. That is the important point; that hasn't been contradicted. There was such a proof that happened in England, that a man could say—Oh, Rhine, that's nothing but guesswork. Exactly; that is guessing, what you call guessing. A hunch is guessing, but a definite guess, you know, is a hunch. That means nothing. You see the point is that it is more than merely probable; it is beyond chance. That's the great problem. But you know, people hate such problems they can't deal with—they can't deal with it. Even Rhine does not understand very often in that respect, because it is a relativation. Now I am going to say something which in these sacred rooms is anathema, a relativation of time and space through the psyche. That's the fact; that is what Rhine has made evident. I'll swallow that. Now that is difficult.

DR. EVANS: May I go a little further, and, of course, some of your recent work, which is indeed very profound, and it is not too well-known to many of our students.

DR. JUNG: Of course not, nobody reads these things, only the general public. Because my books are at least sold.

DR. EVANS: I'm referring to the concept of "synchronicity" which you have discussed, which would have some relevance at this point in our discussion. Would you care to comment a little bit?

DR. JUNG: That is awfully complicated; one wouldn't know where to begin. Of course, this kind of thinking has been started long ago, and when Rhine brought out his results, I thought—now we have at least a more or less dependable basis to argue on, you know. But the argument was not understood at all, because it is really very difficult. You know there are plenty of facts in the observation of the unconscious where you come across cases of a very peculiar kind, of parallel events, namely, that I have a certain thought, a certain definite subject is occupying my attention and my interest. At the same time something else, quite independently, happens that portrays just that thought. This is utter nonsense, you know, looked at from a causal point of view. That it is not nonsense is made evident by the results of Rhine's experiments. There is a probability; it is something more than chance that such a case occurs. I never made statistical experiments except one in the way of Rhine. I have made one for another purpose. But I have come across quite a number of cases where it was most astounding to find that two causal chains happened at the same time, but independent upon each other, so that you could say they had nothing to do with each other. Of course, it's quite clear. For instance, just so, I speak of a red car and at that moment a red car comes here. Now I hadn't seen it; it was impossible because it was behind the building in this moment the red car appears. Now this is an example of mere chance. Yet the Rhine experiment proves that these cases are not mere chances. Of course, many of these things are occurrences where we cannot apply such an argument; otherwise it would be superstitious. We can't say, now this car has appeared because there were some remarks made about a red car. It is a miracle that a red car appears. It is not; it is just chance. But these chances happen more often than chance allows. That shows that there is something behind it. Rhine has a whole institute and with many co-workers and has the means. We have no means here to make such an experiment; otherwise I probably would have done them. But it is just physically impossible. So I have to content myself with the observation of facts.

Third
Interview

Text pages 110–114

DR. EVANS: Dr. Jung, we have been discussing many of the very fundamental ideas in your theories, such as anima, animus, and of course being representative of the whole idea of archetype as you explained it so interestingly, and we talked a little about this whole conception of the ego and the self, the persona, as they are involved in all of this and as we speak, of course, it begins to point in the direction of the sort of question that is so important as we try to understand the individual. This centers around the problem of motivation— why the person does what he does. To a great degree you have already talked about this when you have talked about archetypes and this whole matter. Now in the work of Dr. Freud he spoke of libido as an energy and earlier we spoke of the libido and the fact that you felt it was more than just sexual energy. You thought it could be something much broader. Now you have certain principles about psychic energy which are indeed very provocative. Now one of these principles, I believe you refer to, as the principle of entropy.

DR. JUNG: Well, I only allude to it. The main point, the standpoint of energetics as applied to psychical phenomena. There you have no possibility to measure exactly. So it always remains a sort of analogy. Freud for instance uses the term "libido" in the sense of sexual energy, and that is not quite correct. If it is sexual, then it is a power, like electricity or any other form of manifestation of energy. Now energy is a concept by which you try to express the analogies of all power manifestations; namely, that they have a certain quantity, a certain intensity, and that there is a flow in one direction, namely, to the ultimate suspension of the opposites; low-high-height— a lake on a mountain flows down until all the water is down, you know, then it is finished. And so, you see something similar in the case in psychology. We get tired from intellectual work, or from consciously living, and then we must sleep, you know, to restore our powers, and it is just the same in the night as if the water were pumped from a lower level to a higher level, so that now we can work again the next day. Of course that simile is limping too, so it is only in an analogous way that we use the term "energy," and I used it because I wanted to express the fact that the power manifestation of sexuality is not the only power manifestation, because you have a lot of drives—say the drive to conquer or the drive to aggression, or something of the sort. There are many forms, you see. For instance, take animals, the way they are building their nests, the urge of the traveling bird, you know, that migrates. They all are driven by a sort of energy manifestation, and the meaning of the word "sexuality" would be entirely gone if everything is that, so Freud himself saw that this is not applicable everywhere. And later on he corrected himself by assuming that there are also ego drives. That is something else. That is another manifestation. Now in order not to presume or to prejudice things, I speak simply of energy, a quantity of energy that can manifest via sexuality or via any other instinct. That is the main feature. Not the existence of one single power, because that is not warrantable.

DR. EVANS: Now much of our approach to motivation in our so-called academic psychology today, we speak for example of what we sometimes call a biocentric theory; we say the individual is born with certain inborn physiological, self-preserving types of drives, such as the drive for hunger, thirst, etc. Sex is just one of those, and the satisfaction of these is necessary to maintenance of the organism. Now as the individual comes into the culture, he learns, and these are modified in terms of the society in which he lives. For example, he may, as a result of a particular pattern, develop a specific urge for certain kinds of food, then later he may develop needs for social approval, etc. How would this statement which is the sort of thing we often observe as we try to understand motivation; would this be consistent with your ideas? Then you would say that these basic, innate, instinctual patterns can be modified by the environment—by the culture?

DR. JUNG: Yes, certainly.

DR. EVANS: Now another thing about motivation, or the condition which arouses, directs and sustains the individual; there seems to be two views that we find in much of our psychology in America today. One might be called an historical view, where we try to look in the history and development of the individual for answers as to why he is doing a certain thing at the moment. Then we have another view which was postulated and discussed by Dr. Kurt Lewin, which is a sort of a field theory. He thinks that the history—the past—is not important in motivation, but all conditions which affect the individual at that moment. We can predict behavior by knowing all the conditions of that moment and we don't have to go back to the past to understand why the person does what he does. Do you think that the present field idea of Dr. Lewin has any virtue? . . .

DR. JUNG: Well, obviously I always insist that even a chronic neurosis has its true cause in the moment—now. You see, neurosis is made every day by the wrong attitude the individual has, but that wrong attitude is a sort of fact, and needs to be explained historically by things that have happened in the past. But that is one-sided too, because all psychological facts are oriented, not only to a cause, but also to a certain goal. They are, in a way, teleological, namely, they serve a certain purpose. And so the wrong attitude can have originated in a certain way long ago, but it wouldn't exist today anymore if there were not immediate causes and immediate purposes that keep it alive today. And so a neurosis can be finished suddenly on a certain day, in spite of all causes. Further, one has observed, in the beginning of the war, cases of compulsion neurosis that have lasted for many years, and suddenly they were cured because they got into an entirely new condition. It is like a shock, you see. Even schizophrenia can be vastly improved by a shock because that's a new condition; it is a very shocking thing, that shocks them out of their habitual attitude; they are no more in it, and then the whole thing collapses—the whole system that has been built up for years.

Text pages 122–124

DR. EVANS: So in therapy, in working with a patient, you would not say that it is absolutely imperative to have to reformulate all of his past life in order to help him with his present neurosis? You feel that you could deal with him as of the moment—with his problem as it is at this time, and it is not necessary to go back and probe into things that happened to him during his second or third year of life.

DR. JUNG: You see, there is no system about it in therapy. In therapy you treat the patient as he is in the present moment, irrespective of causes and such things. That is all more or less theoretical. There are cases who know just as much about their own neurosis as I know about it, in a way. In such cases I can start right away with posing the problem. For instance, there is a case—a professor of philosophy—and he imagines that he has cancer. He shows me several dozen X-ray plates that prove there is no cancer. He says, "Of course I have no cancer, but nevertheless I'm afraid I could have one; I've consulted so many surgeons and they all assure me there is none, and I know there is none, but I might have one." You see? And that's enough. Such a case can stop from one moment to the other; the sickness stops thinking such a foolish thing. But that is exactly what he cannot do. And in such a case I say, "Well, it's perfectly plain to you that it's nonsense what you believe. Now, why are you forced to believe such nonsense? What is the power that makes you think such a thing against your free will? You know it is nonsense, and why should you think it? or what forces you to think it? What is that power that makes you think such a thing? It's like a possession, you know. Exactly. Like a demon in him, that makes him think like that, in spite of the fact that he doesn't want it. That is the problem for an intellectual man. Then I say, "Now you have no answer; I have no answer. Now what are we going to do?" I say, "Now we will see what you dream," because a dream is a manifestation of the unconscious side. Now he never has heard of the unconscious side, so I must explain to him that he has an unconscious and that the dream is a manifestation of it, and if we succeed in analyzing the dream we might get an idea about that power that makes him think like that. So, in such a case one can begin right away with the analysis of dreams. And in all cases that are a bit serious (mind you, this is not a simple case; this is a very serious and difficult case, in spite of the simplicity of the phenomenology, of the symptomatology);—in all cases after the preliminaries, as it were—the history of the family, the whole medical analysis, etc.—we come to that question, what is it in your unconscious that makes you wrong, that hinders you from thinking normally? Then we are there where we can begin with the observation of the unconscious. Then, day-by-day, one goes on by the data the unconscious produces. You see, we discuss the dream and that gives a new surface to the whole problem; and he will have another dream, and the next dream gives again an answer, because the unconscious is in a compensated relation to consciousness, and after a while we get the full picture. And if he has the full picture and has the necessary moral stamina, then he can be cured. But in the end it is a moral question, whether a man applies what he has learned or not.

Text pages 75–79

DR. EVANS: Then in a sense, in this situation the unconscious plays a very important part—the unconscious which is found in the dream—but as you see it, what you have found in this dream is not necessarily then an image or symbol of what has happened in the past in his particular case.

DR. JUNG: Oh no! It just is a symbol of the—symbol, you see, that is a special term; it is the manifestation of the situation of the unconscious, looked at from the unconscious. You see, I tell you, for instance, something which is my personal subjective view. And if I ask myself, "Now, are you really quite convinced of it?" well, I must admit I have certain doubts. There are certain doubts, not in the moment when I tell you, but these doubts are in the unconscious. And when I have a dream about it, these doubts come to the forefront in my dreams. And that is the way the unconscious looks at the thing; it is just as though the unconscious says, "It is all very well what you are stating, *but* you omit entirely such and such a point."

DR. EVANS: Now, if the unconscious acts on the present situation—looking at this in broad motivational terms—this effect of the unconscious is not something which is a result of repression in the way the orthodox psychoanalyst looks at it at all, then——

DR. JUNG: It may be, you know, that what the unconscious has to say is so disagreeable that one prefers not to listen. And in most cases people would be probably less neurotic if they could admit the things; but these things are always a bit difficult or disagreeable, inconvenient, or something of the sort. So there is always a certain amount of repression, but that is not the main thing. The main thing is that they are really unconscious. If you are unconscious about certain things that ought to be conscious, then you are dissociated. Then you are a man whose left hand never knows what the right is doing, and counteracts or interferes with the right hand. Now such a man is hampered all over the place.

DR. EVANS: Now, looking at the unconscious in this way; of course, as you say, if it's unconscious, how do we know about it; but trying for the moment just as an illustration— with a particular individual, we might say, who has been brought up in a culture such as the culture of India. This particular individual in India, if we could examine his unconscious, would it be in many respects similar to the unconscious of a particular individual who, we'll say, lived in Switzerland all his life? You spoke earlier about these universals. Would there be a great deal of equivalence between the unconscious of a particular individual who was raised in one culture, and another individual who was raised in an entirely different culture?

DR. JUNG: Well, that question is also complicated, because when we speak of the unconscious, we almost should say "which unconscious?" Namely, is it that personal unconsciousness which is characteristic for a certain person, for a certain individual?

DR. EVANS: You have the personal unconsciousness; this is one kind of unconscious.

DR. JUNG: In treatment, for instance, the treatment of neurosis, you have to do with that personal unconsciousness for quite a while, and then only dreams come that show that the collective unconscious is touched upon. As long as there is material of a personal nature you have to deal with the personal unconscious. But when you get, say, to a question, to a problem which is no more merely personal but also collective, you get collective dreams.

DR. EVANS: Now the distinction between the personal unconscious and the collective unconscious, then, is that the personal could be more involved with the immediate life of the individual, and the collective would be universal—it would be the same elements in all men?

DR. JUNG: It would be collective. For instance, in all of us the psyche has collective problems—collective convictions, etc. We are very much influenced by them. For instance, you belong to a certain political party, or to a certain confession;

that can be a serious determinant of your behavior. Now, as a matter of personal conflict doesn't touch upon it, it's no question, it doesn't appear; but the moment you transcend your personal sphere and come to your own personal determinants—say to a political question, or any other social question which really matters to you—then you are confronted with a collective problem, and then you have collective dreams.

Text pages 135–138

DR. EVANS: I wonder if it would be too presumptuous of me to ask if you could for the moment think of an actual case of perhaps a patient or a friend in which you might show us, that is, specifically, how the personal and the collective unconscious were acting in neurosis or perhaps in a problem that he may have had.

DR. JUNG: Wait a moment, there is an enormous amount of personal dreams, for instance, and I couldn't possibly tell you—but in research there are millions of such personal dreams that simply deal, for instance, with the fact how your relation is to your father or to your mother or to your wife, etc., with all sorts of individual variations. But suppose a patient comes to that level, or that his conflict begins to become really very serious so that his mind might suffer. Then he can have a collective dream in which really mythological motifs appear. There are plenty of examples in literature. I wrote a book, you know, an introduction to the psychology of the unconscious, about such dreams. I remember, for instance, a case; it was a very learned man, and very rational. He had of course a lot of personal problems, but they got so bad that he got into very disagreeable relations to his whole surroundings. He was a member of a society and he got into a brawl with the people of that society; it was really quite shocking. Now, he started with collective dreams. Suddenly, he dreamt of things he had never thought of in his life before; mythological motifs, and he thought he was crazy, because he couldn't understand it at all—just as if the whole world were transformed. That is what you see in cases of schizophrenia, but that is not a case of schizophrenia. As examples you can take any of these collective dreams I have published—there are plenty. For the moment I cannot remember a suitable example. To make it clear I should tell a long story, and then you'll see where it applies. Otherwise it makes no sense to have something short. I told you the case of that intuitive girl who suddenly came out with the statement that she had a black snake in the belly. Well now, that is a collective symbol. That is not an individual fantasy—that is a collective fantasy that is well-known in India. For instance, she had nothing to do with India; but we have it too, we are similarly human. But it is entirely unknown. So that I even, in the first moment, thought, perhaps she's crazy. But she was only highly intuitive. It is in India known, it is at the basis of a whole philosophical system, of Tantrism. This is Kundalini, Kundalini the serpent. And that is something known to some few specialists: generally it is not known that we have a serpent in the abdomen. But that is a collective dream or collective fantasy.

DR. EVANS: In day-to-day living of the individual, is it possible that things that trouble them and cause tension that he represses—now these things that he represses will become part of the personal unconscious?

DR. JUNG: Yes. He doesn't repress consciously always. These things disappear, and Freud explains that by active repression. But you can prove that these things never have been conscious before. They simply don't appear, and you don't know why they don't appear. Of course, *après tout,* you can say, when it comes up—that is why they didn't appear, because they were disagreeable or incompatible with his conscious views, with his conscious attitude; but that is afterwards; you couldn't predict it. So you see, these things that have an emotional tone, they are partially autonomous; so that they can appear or the contrary, they do not appear. They can disappear at wish, not of the subject, but of their own, or you also can repress them. It is so the same with projections. For instance, people say, "one makes projections." That's nonsense. One doesn't make them; one finds them. They are already there, because the unconscious is here not conscious, but there it is conscious; in my brother; I see the beam of my eye suspended in my brother's eye. That is right there because I am unconscious of the beam in my eye. Projections are not made. And so these disappearances, or so-called repressions, are just like projections. Instead of being projected into somebody or into something outside, they are introjected; they are already unconscious. But you are not the one that is doing it. There are cases, sure, but I should say the majority of the cases aren't repressions. That was my first point of difference with Freud. I saw in the association experiment that certain complexes are quite certainly not repressed. They simply won't appear. Because you see the unconscious is real; it is an entity; it works by itself, it is autonomous.

DR. EVANS: So in a sense, and looking at the so-called defense mechanisms which the orthodox psychoanalysts speak of: projection, rationalization, etc., then in a sense the way you would differ from the orthodox psychoanalytic view— you would not say that they are a repressed type or way of defending the manner in which the ego is being defended. Rather you would say they are already there; it is simply a manifestation of patterns that are already present in the unconscious.

DR. JUNG: Yes. For instance, take the example of that serpent. That never had been repressed, otherwise it should have been conscious to her; it was on the contrary unconscious to her and it only appeared in her fantasy. It appeared spontaneously. She didn't know how she came to it. She said, "Well, I just saw it."

DR. EVANS: Now some of the orthodox psychoanalysts might have said—this is a phallic symbol.

DR. JUNG: But you can say anything; you know we can say a church spire is a phallic symbol, but when you dream of a penis—what is that? You know what an analyst said, one of the orthodox—the old guard—he said, "In this case the censor has not functioned." You call that a scientific explanation?

Text pages 114–115

DR. EVANS: You have brought up very many interesting and provocative ideas here, as you have of course in your many, many thousands of pages of writing, and of course running through all your work there are many of these ideas; the personal unconscious, the race unconscious, the self, the ego, the persona, the energy principle such as we mentioned, entropy and equivalence; in a sense the first and second law of thermodynamics, I believe, that you suggest or allude to. Now in trying to look at the whole person (and one is struck that you are trying hard to look at the whole person—that you don't want to look at the little parts of the person, but feel that sort of a total future realization of the whole that may be very important). You talk, for example, about a process of individuation—how this whole person emerges. Would you like to comment a little bit upon this process of individuation, how all these factors move toward a whole—a totality?

DR. JUNG: Well, you know, that's something quite simple. Take an acorn, put it into the ground, it grows and becomes an oak. That is man. Man develops from an egg, and develops into the whole man, and that is the law that is in him.

DR. EVANS: So you think the psychic development is in many ways like the biological development.

DR. JUNG: The psychic development is out of the world —it is not something else; or an opinion. It is a fact that people develop in their psychical development on the same principle as they develop in the body. Why should we assume that is a different principle? It is really the same kind of evolutionary behavior as the body shows. Take for instance, those animals that have specially differentiated anatomical characteristics, those of the teeth, or something like that—well, they are in accordance with their mental behavior, or the mental behavior is in accordance with those organs.

DR. EVANS: So as you see it, then there is no need to bring in other types of ideas, other types of theories to explain development—the basic biological law is still——

DR. JUNG: The psyche is nothing different from the living being. It is the psychical aspect of the living being. It is even the psychical aspect of matter. It is a quality.

DR. EVANS: Now in terms of this total growth——

DR. JUNG: Not the faintest trace of psychological, so I couldn't discuss the psychological implications.

DR. EVANS: Well, you, yourself, have had a lot of background in physics and you know——

DR. JUNG: No, nothing to speak of.

Text pages 155–158

DR. EVANS: As one reads your work we seem to be aware that you know archeology, anthropology——

DR. JUNG: Well, inasmuch as much of my work is concerned in it, but I have no mathematical gifts, you know. You cannot get real knowledge, you know, or understanding of nuclear physics without a good deal of mathematics, higher mathematics. There I only have a certain relation with it in the epistemological questions, you know. Modern physics is truly entering the sphere of the invisible and intangible, as it were, and everything is really conclusion. It is in reality a field of probabilities, you know, that is exactly the same as the unconscious. Therefore I often have discussed this with a professor who is a nuclear physicist, and to my amazement I found that they have terms which are used in psychology too, and simply on account of the fact we are entering a sphere—the one from without, and we from within, which is unknown. That's the reason for the parleys—the sort of parleys with higher mathematics. For instance, we use the term transcendental function. Now transcendental function is a mathematical term, the function of irrational and imaginary numbers. Now that is higher mathematics, with which I have nothing to do. But we come to the same terminology.

DR. EVANS: When you spoke with Dr. Einstein in your early discussion, he more or less tried out some of his ideas on you. Did you ever bring to him the possibility that relativity might apply to psychic functions? Did you ever discuss that?

DR. JUNG: Well, you see, you know how it is when a man is so concentrated upon his own ideas, and when he is a mathematician on top of everything, then you are not welcome.

DR. EVANS: What year was it that you were friends with Einstein?

DR. JUNG: I wouldn't call myself a friend, I was simply the host. I tried to listen and to understand; and so there was little chance to insert some of my own ideas.

DR. EVANS: Was this after he had already formulated his relativity theory? Or just before?

DR. JUNG: That was when he was working on it, right in the beginning. It was very interesting.

DR. EVANS: In your dealings with Professor Toynbee—have you gotten rather interested in his ideas of history?

DR. JUNG: Ah, yes. His ideas about the life of civilizations that is, they are ruled by archetypal forms, and Toynbee has seen what I mean by the historical function of archetypal developments. That is a mighty important determinant of human behavior, that lasts for centuries or thousands of years, and it expresses itself in symbols, sometimes symbols, you would think nothing of it. You know that Russia, the Soviet Republic, had that symbol of the red star. Now it is a five-rayed red star. America has the five-rayed white star. They are enemies; they can't come together. In the Middle Ages, for about at least 200 years, the red and the white is the couple that is ultimately destined to marry each other. So America is a sort of matriarchy, as much as most of the money is in the hands of women. Russia is the land of the little father; that's a patriarchy. So it is father and mother; the white woman; in the Middle Ages they called the white woman *"femina candida"*; the *"servus rubeus,"* the red slave. The two lovers have quarreled with each other.

Text pages 85–87

DR. EVANS: Now after this very pleasant little digression from the point, and I think these digressions are most enjoyable, we are trying to get for our students clarification of some of these points. Just before we began discussing physics, your theories, Einstein, Toynbee, etc., I was about to discuss or ask you a little bit about what seems to be a very fundamental part of your writing and ideas, and it is encased in the term "mandala." Now that seems to be a sort of ultimate realization or direction, and I would certainly be most interested in hearing your observations on this idea.

DR. JUNG: Mandala is just one typical archetypal form; it is what they called in alchemy the *quadratura circuli*— the square in the circle, or the circle in the square. It is an age-old symbol that goes right back to the prehistory of man. It is all over the earth and it either expresses the Deity or the self, and these two terms are psychologically very much related, which doesn't mean that I believe that God is the self or that the self is God. I made that statement that there is a psychological relation. There is plenty of evidence for it. It is a very important archetype—it is the archetype of inner order, and it is always used in that sense, either to make arrangements of the many, many aspects of the universe, a world scheme, or the scheme of our psyche, and it expresses the fact that there is a center and a periphery and it tries to embrace the whole. It is the symbol of wholeness. So you see, in a moment where, during the treatment, when there is a great disorder and chaos in a man's mind, then the symbol can appear, as in the form of a mandala in a dream, or when he makes imaginary fantastical drawings, or something of the sort. Then it spontaneously appears as a compensatory archetype, bringing order, showing the possibility of order, and it means a center which is not coincident with the ego, but with the wholeness—it is wholeness—the wholeness which I call the self; this is the term for wholeness—and I am not whole in my ego; my ego is a fragment of my personality. So you see, the center of a mandala is not the ego; it is the whole personality; the center of the whole personality, and it plays a very great role in the East, for instance, but in the Middle Ages equally. Then it has been lost, and it has been thought of as a mere sort of allegorical decorative motif; but as a matter of fact it is highly important and highly autonomous, a symbol that appears in dreams etc., or in folklore. We could say that it is the main archetype.

Text pages 119–122

DR. EVANS: In speaking of this totality which you say is a sort of unified self, and of course as you suggest here, that the mandala is a very important archetype symbolizing this balance that you alluded to here. Of course sometimes we try in psychology to start from whatever totality does exist in the individual and try to look into underlying motivation, and we have recently used a great deal of testing which we call "projective tests." As you have already suggested, and we all know of course very well, that you certainly played a major role in developing this point of view with your word association method. Of course, to a great degree, another very interesting development, in 1922 Hermann Rorschach developed his inkblot, or Rorschach test. Now first of all, to even talk a little bit more about word association—we've talked a little about it already—would you like to get into it in a little bit more detail? First of all, what are the ingredients of the word association test? What is involved in the use of it?

DR. JUNG: You mean the practical use of it?

DR. EVANS: Yes.

DR. JUNG: Oh, well, you see, in the beginning when I was a young man, of course I was completely disoriented with patients. I didn't know where to begin or what to say, and the association experiment has given me a chance of access to their unconscious. I learned about the things they did not tell me, and I got a deep insight into the things that they did not know, and I discovered many things.

DR. EVANS: Now would you say that from responses, you discovered complexes, or sort of emotional blocks; of course this word "complex" originated with you and it's used very widely now.

DR. JUNG: Yes, the *Gefühlsbetonter Komplex,* the "Complex"; that is one of my terms.

DR. EVANS: How do you hope from these complexes or emotional blocks that you uncovered as you administer this test, to get at materials of the personal unconscious or the racial unconscious, or all of the factors——

DR. JUNG: You know, in the beginning there was no question of collective unconscious or something like that. It is chiefly the ordinary personal complexes.

DR. EVANS: I see; you weren't expecting to get into such depths.

DR. JUNG: Among hundreds of complex associations there might appear an archetypal element, but that wouldn't show particularly. That is not the point. You know it is like the Rorschach, a superficial orientation.

DR. EVANS: You knew Hermann Rorschach, I believe, did you not?

DR. JUNG: No, he has circumvented me as much as possible.

DR. EVANS: But did you get to know him personally?

DR. JUNG: No, I never saw him.

DR. EVANS: In his introtensive and extrotensive ideas, of course, he is talking a great deal about your introversion and extroversion ideas, in his own interpretation.

DR. JUNG: Yes, but I was anathema because I had first said it and that is unforgivable. I never should have done it.

DR. EVANS: So you really didn't have any personal contacts with Rorschach?

DR. JUNG: No personal relations at all.

DR. EVANS: Are you familiar with his test? Have you seen his test?—the Rorschach test?

DR. JUNG: Yes, but I never applied it because later on I didn't apply the association test anymore, because it wasn't necessary; I learned what I had to learn but from the exact examination of psychic reactions. I think it is a very excellent means.

DR. EVANS: Would you say, for example, that the practicing psychiatrist, the clinical psychologist and practicing psychiatrists, could use these projective tests like the Rorschach test?

DR. JUNG: For instance you see, too, for education of psychologists, practical psychologists meant to do actual work with people, I think it was the best means to make them see how the unconscious works.

DR. EVANS: The projective tests——

DR. JUNG: It is exceedingly didactic. There one can demonstrate repression or the amnestic phenomenon, the way in which people cover their emotions, etc. It is like an ordinary conversation, but seen and measured in its principles. That makes it so interesting. You observe all the things you observe in a conversation. For instance, you ask people something and discuss certain things, and you observe little hesitations, mistakes in speech, all that comes to the foreground, and they are measurable, you know, in the experiment. And so I think I don't overrate the didactic value of it; I think very highly of it. And still you use it in education of young analysts. Or, if I have a case that doesn't want to talk I can make such experiment and find out a lot of things through the experiment. I have, for instance, discovered a murder.

DR. EVANS: Is that right? Would you like to tell us how this is done.

DR. JUNG: You see, you have that lie detector in the United States.

DR. EVANS: Yes.

DR. JUNG: Well, that's an association test I have worked out with the psycho-galvanic phenomenon, and we have done a lot of work also on the pneumograph, to show the decrease of the volume of breathing under the influence of a complex. That's one of the reasons of tuberculosis—that rises from such a condition; people have a very shallow breathing, don't ventilate the apices of their lungs anymore, and get tuberculosis. Half of the tubercular cases are psychic.

Text pages 143–149

DR. EVANS: This matter of psychic tubercular cases etc., gets us into a particularly interesting special case of motivation that we're talking a lot about in the United States, and I'm sure is of interest to you, as this is the whole area of psychosomatic medicine. The way our unconscious, our emotional elements of the total personality——

DR. JUNG: I see a lot of astounding cures of tuberculosis —chronic tuberculosis, by an analyst, where people learn to breathe again. It would not help them if they had learned to breathe normally, that would not have helped. But the understanding—what their complexes were—that has helped them . . .

DR. EVANS: When did you first become interested in the psychic factors of tuberculosis? Many years ago?

DR. JUNG: I was an analyst to begin with; I was always interested. Naturally. Because I understood so little of it. Or I noticed that they understood so little.

DR. EVANS: The point that this leads to very logically— the point I started making a moment ago—this is very much the heart of a lot of our thinking, the area of what we call psychosomatic medicine. We are right now becoming more and more interested in the very thing you are saying—how the emotional, how the unconscious, how these personality factors can actually have an effect on the body. And of course the classic example in the United States is the peptic ulcer. We believe this is a case where emotional factors have actually created the pathology. And we have extended these ideas into many other areas. We, for example, feel that where there already is pathology, these emotional factors can intensify it; not only can create it, but can intensify existing pathology. Or sometimes there may be actually symptoms of the pathology without these patterns. For example, many of our physicians, or as you say analysts, in America say that sixty to seventy percent of their patients are not individuals who have anything really physically wrong with them, but they have disorders of psychosomatic origin.

DR. JUNG: Yes, that is well-known—since more than fifty years. But the question is, how to cure them.

DR. EVANS: Speaking of these psychosomatic disturbances, and as you say tuberculosis would be one example, and we are now interested in further research in this matter in the United States, and for that matter other parts of the world, do you have any ideas as to why the patient selects this type of symptom? Why does he show——

DR. JUNG: He doesn't select; they happen to him. You could ask just as well when you are eaten by a crocodile, how you happened to select that crocodile; he has selected you.

DR. EVANS: Of course it means that in a sense, unconsciously, why you selected——

DR. JUNG: No, not even unconsciously.

DR. EVANS: You don't believe there is any way of tracing in the personality of the individual the reasons why——

DR. JUNG: That is an extraordinary exaggeration of the importance of the subject, as if he were choosing such things. They get him.

DR. EVANS: Even unconsciously there isn't this degree of freedom—now in the whole matter of psychosomatic medicine, they have now for example, very recently hinted that cancer may have psychosomatic involvements.

DR. JUNG: Yes, yes.

DR. EVANS: And this doesn't surprise you?

DR. JUNG: Not at all. We know these since long ago, you know; fifty years ago we already had these cases; ulcer of the stomach, tuberculosis, chronic arthritis, skin diseases; all are psychogenic under certain conditions.

DR. EVANS: And even cancer—the fact that cancer may have——

DR. JUNG: Well, you see, I couldn't swear, but I have seen cases where I thought, I should wonder whether there was not a psychogenic reason for that particular ailment; it came too conveniently.

DR. EVANS: And some of these studies, for example, in the United States show that Jewish women practically never get cancer in the vaginal regions, but more often get cancer say in the breast region.

DR. JUNG: Well, many things can be found out about cancer, I'm sure. You see to us it was always a question of how to treat these things, and anything is possible. Every disease has a psychological accompaniment, and it all depends—perhaps life depends—upon it, whether you treat such a patient psychologically in the proper way or not. That can help tremendously. Even if you cannot prove in the least that the disease in itself is psychogenic. Or you can have an infectious disease in a certain moment of a psychical ailment or predicament, because you are particularly accessible to an infection. Tonsilitis is such a typical psychological disease; yet it is not psychological in its physical causation. It's just an infection. But why, then—well, it was the psychological moment. When it is established and there is a high fever and an abscess, you cannot cure it by psychology. Yet it is quite possible that you can avoid it by a proper psychological attitude.

DR. EVANS: So all this interest in psychosomatic medicine is pretty old stuff to you.

DR. JUNG: These things were all known here long ago.

DR. EVANS: And you are not at all surprised at the new developments——

DR. JUNG: For instance, there is the toxic aspect of schizophrenia. I published it fifty years ago—just fifty years ago, and now everyone discovers it. You are far ahead in America with technological things, but in psychological matters and such things you are fifty years back. You simply don't understand them; that's a fact. I don't want to offend you, that's a general corrective statement; you simply are not yet aware of what there is. There are plenty more things than people have an idea of. I told you that case of the philosopher who didn't even know what the unconscious was; he thought it was an apparition. Everyone who says that I am a mystic is just an idiot. He just doesn't understand the first word of psychology.

DR. EVANS: There is certainly nothing mystical about the statements you have just been making. Now to pursue this a little further, another development that falls right in line with this whole discussion of psychosomatic medicine has been the use of (of course historically drugs have been used a great deal by people to try to forget their troubles, to relieve pain, etc.); but a particular development has been the so-called non-addictive drugs which of course began in France with the drug chlorpromazine, then Reserpine, Serpentina, and a great variety of lesser drugs which are known by such trade names as Miltown, etc., which are now being administered very freely to patients both by the physician, the general practitioner, the internist; and it is not only drugs that are being administered in the stronger forms to the mentally ill patients—the schizophrenics and others who are not approachable, but also to a tremendous degree today particularly in America we know that these drugs are being dispensed almost as freely as aspirins now to reduce the tensions.

DR. JUNG: It is very dangerous.

DR. EVANS: Why do you think this is dangerous? They say these are nonaddictive.

DR. JUNG: It's just like the compulsion that is caused by morphine, or heroin. It becomes a habit. You don't know what you do, you see, when you use such drugs. It is like the abuse of narcotics.

DR. EVANS: But the argument is that these are not habit-forming; they are not addictive, not physiologically.

DR. JUNG: Oh, yes; that's what one says.

DR. EVANS: But you feel that psychologically there is still addiction?

DR. JUNG: Yes, for instance, there are many drugs that don't produce habits like morphine, yet it becomes a psychical habit, and that is just as bad as anything else.

DR. EVANS: Have you actually seen any patients or had any contact with individuals who have been taking these particular drugs, these tranquilizers?

DR. JUNG: I can't say. You see with us there are very few. In America, you know, there are all the little powders and the tablets. Happily enough we are not yet so far.

DR. EVANS: So you feel that, even though there may not be any physiological basis for habit-forming aspects of these drugs—that psychologically——

DR. JUNG: You see, American life is, in a subtle way, so one-sided and so uprooted, that you must have something to compensate the earth. You see, you have to pacify your unconscious all along the line because it is in absolute uproar, so at the slightest provocation you have a big moral rebellion in America. Look at the rebellion of modern youth in America, the sexual rebellion, and all that. Because the real natural man is just in open rebellion against the utterly inhuman form of life. They are absolutely divorced, you know, from nature, in a way, and that accounts for that drug abuse.

DR. EVANS: But on the other hand, in the treatment of serious mental patients, for example we have the problem of hospitalized patients, the schizophrenics, manic-depressives —these individuals. For instance certain schizophrenics are withdrawn and you can't deal with them in therapy. Now in many of the hospitals in the United States they have been using some of these drugs such as chlorpromazine, and the patient comes back to reality for a short time, and thereby becomes amenable to therapy. Now you take the patients who can never be approached by therapy—I don't think most of our practitioners believe the drugs cure the patients in themselves, but they at least make the patient amenable to therapy.

DR. JUNG: Yes; the only question is whether that amenability is a real thing or drug-induced. I am sure that any kind of suggestive treatment will have effect, because they simply become suggestible. Because you see, any drug or shock in the minds—they lower stamina, and these people become accessible to suggestion, and they can be led, can be made into something, but it is not a very happy result.

Fourth Interview

DR. EVANS: Professor Jung, in our several interesting discussions so far, a great deal of our talk has been on more or less some of the very fundamental aspects of your theories and writings, and of course our main purpose in getting some of your reactions to these fundamental ideas is so that our students, many of whom have not had an opportunity to study a great deal of your work, might be introduced to some of your ideas from you personally, which of course is probably the best way they possibly could be. Now of course, as one looks over your work, one is impressed with a much broader scope than just the beginning fundamental outline of a theory of personality. You have investigated a much broader problem, and of course one of the things that I feel you could comment on very, very well and would be of great interest to us, centers around the basic problem of the kind of training—the kind of background a psychologist, a person who wants to study the individual, should have. For example, there is one view that says maybe he should be trained only in statistics and rather rigorous scientific method, and that this is the greatest tool he can have. However, there may be more to this problem of studying the individual than this rather narrow type of training. Would you like to comment on the type of training, the kind of background you think should be required of an individual as he tries to understand and study the human organism?

DR. JUNG: I don't quite understand what you mean.

Text pages 149–152

DR. EVANS: For example, just to be a little bit more specific, do you think that the humanities are important for the individual who wants to study the individual?

DR. JUNG: Well, of course when you study human psychology you can't help noticing that man's psychology doesn't only consist of the ramifications of instinct in his behavior; these are not the only determinants, there are many others, and the study of man from his biological aspect only is by far insufficient. To understand human psychology it is absolutely necessary that you study man; also in his social and general environments. And so you have to consider, for instance, the fact that there are different kinds of societies, different kinds of nations, different traditions, and to that purpose it is absolutely necessary that one treats the problem of the human psyche from many standpoints. Each is a most considerable task, naturally. Thus, when I—after my association experiments —when I realized that there is obviously an unconscious; the question holds, now what is this unconscious? Does it consist merely of rests—of remnants of conscious activities, or are there things that are practically forever unconscious? In other words, is the unconscious a factor in itself—and I soon came to the conclusion that the unconscious must be a factor in itself, because I observed time and again, for instance, in people's dreams or schizophrenic patients' delusions, fantasies, contain the motifs which they couldn't possibly have acquired in our surroundings. This depends upon the fact that the child is not born *tabula rasa,* but it is a definite mixture or combination of genes, and although the genes seem to contain chiefly dynamic factors (*arrangeurs*) of certain behavior, they have a tremendous importance, also for the arrangement of the psyche. The psyche too, inasmuch as it appears naturally. Before this appears you cannot study it, but as much as it appears, it has certain qualities; it has a certain character, and that needs must depend upon the elements born in the child. So, factors determining human behavior are born with the child, and determine its further development. Now that is one side of the picture. The other side of the picture is, the individual lives in connection with others in certain definite surroundings that will influence the given combination of qualities. And that now is also a very complicated factor, because the environmental influences are not merely personal. There are any amount of objective factors. The general social conditions, laws, convictions, ways of looking at things, of dealing with things. These things are not of an arbitrary character. They are historical. There are historical reasons why things are as they are. And as there are historical reasons for the qualities of the psyche, that it's formed, there is such a thing as the history of man's evolution in past aeons, and that shows that real understanding of the psyche must consist in the elucidation of the history of the human race. History of the mind, for instance, as of the biological data.

Text pages 132–135

So, when I wrote my first book concerning the psychology of the unconscious, I already had formed a certain idea of the nature of the unconscious. To me it was then a living remnant of the original history of man, living in his surroundings—a very complicated picture. And my material then, my empirical material, was formed chiefly by lunatics—by cases of schizophrenia, and there I had observed that there are, chiefly in the beginning of a disease, invasions of fantasies into conscious life, fantasies of an entirely unexpected sort, most bewildering to the patient. He gets quite confused by these ideas and he gets into a sort of panic because he never before had thought such things; they are quite strange to him, and equally strange to his physician. You see the analyst is equally dumbfounded by the peculiar character of those fantasies. Therefore, one says, "That man is crazy. He is crazy to think such things; nobody thinks such things," and the patient agrees with him. He would at least agree or he does even agree. And all the more he gets into a panic. So, as an analyst I thought it to be really the task for psychiatry to elucidate that thing that broke into consciousness. These voices, these delusions. In those days —that is, mind you, more than forty years ago, over fifty years ago—I had no hope to be able to treat these cases or to be able to help them, but I had a great scientific curiosity, and I wanted to know what these things really were, because I thought that these things had a system, that they were not merely chaotic, decayed material, because there was too much sense in those fantasies. So, what I did then was—I studied

cases of psychogenic diseases, like hysteria, hysterical somnambulism, and such things, where the contents that came from
the unconscious were in a readable condition—an understandable condition, and then I saw that, in contradistinction to the
schizophrenics, the mental contents of hysterics were of a
humanly understandable character, there were even elaborate,
dramatic, suggestive, insinuating, so one could make out a
second personality. Now this is not the case in schizophrenia.
There the fantasies, on the contrary, are unsystematic, chaotic,
and you cannot make out a proper second personality, apart
from rare cases of a complicated nature. Now, I knew of
psychopathic cases, in the bounds between schizophrenia and
hysteria, where ideas came up, delusions that were not exactly
hysterical, because they were singularly difficult to understand,
sort of strange eruptions, and I thought that these cases could
give me a better understanding, so I took the opportunity when
Professor Flournoy, the old professor of psychology and
philosophy at the University of Geneva, when he published a
case of an American girl who had bestowed upon him a series
of half poetic and romantic fantasies. He published that material without commenting on them. He gave it as an example
of creative imagination. Now, when I read those fantasies I
saw this is exactly the material, and I was always a bit afraid
that people, when I tell of my personal experiences with
patients, that they will say, this is merely suggestion, you
know. I took that case because I surely had no hand in it.
It was old Professor Flournoy, an authority, he was a friend
of William James. I knew him personally very well—a fine
old man, and he certainly wouldn't be accused of having influenced the patient. That is the reason why I analyzed these
fantasies. That case became the object of a whole book, called
The Psychology of the Unconscious. It is called now,
Symbols of Transformation. And I have revised it after
forty years. It needed it, because it was the first attempt.
And there I tried to show that there is a sort of unconscious—
I then simply called it "the unconscious"—that clearly produces
things which are historical and not personal. It was mythological material, the nature of which was not understood,
either by Professor Flournoy or by the patient. And there
I tried for the first time to produce a picture of the functioning
of the unconscious—a functioning which allowed certain conclusions as to the nature of the unconscious. Then, after I had
written that book, that cost me my friendship with Freud
because he couldn't accept it; to him the unconscious was a
product of consciousness, and simply contained the remnants;
I mean it was a sort of storeroom where all the discarded things

of conscious were heaped up and left. To me the unconscious, then, was already a matrix, a sort of basis of consciousness of a creative nature, namely, capable of autonomous acts—autonomous intrusions into the consciousness. In other words, I took the existence of the unconscious for a real fact—for a real autonomous factor, capable of independent action. That was a psychological problem of the very first order, and that made me think and feel—because the whole of philosophy in our days has not yet recognized this fact—that we have a counter factor in our unconscious; that in our psyche there are two; consciousness is one factor, and there is another factor, equally important, that is the unconscious, that can interfere with consciousness any time it pleases. And of course, I say to myself now—this is very uncomfortable, because I think I am the only master in my house, but I must admit that there is another—somebody in that house that can play tricks, and I have to deal with the unfortunate victims of that interference every day in my patients. So the next thing I wrote was in 1916, namely, a disquisition about the relation between the ego and the unconscious.

Text pages 94–95

There I tried to formulate the experiences that are more or less regularly observable in cases where consciousness is exposed to unconscious datas, to interferences or intrusions, where the unconscious is considered as an autonomous factor, that has to be taken seriously, where one doesn't say anymore or undervalue the unconscious by assuming that it is nothing but a discarded remnant of consciousness. It is a factor in its own dignity, and a very important factor, because it can create the most horrible disturbances. That was a pamphlet I wrote; it was published in French, and nobody understood it. I saw that the reason why nobody understands it is that why nobody has had a similar experience, because the question hasn't been pursued to such an end, namely, that one has taken the unconscious seriously and considered it as a real factor that can determine human behavior to a very considerable degree. So, I began an examination of the human attitudes, namely, how our consciousness functions. I couldn't help see, for instance, the difference between Freud and Adler; typical difference. The one assumes that the things evolve along the line of the sex instinct. The other assumes that things evolve along the line of the power drive. And there I was—in between the two. I could see the justification of Freud's view, and also could see the same for Adler, and I knew that there are plenty of other ways in which things could be envisaged. And so I considered it my scientific duty to examine first the condition of the human consciousness. That is the originator of ways of envisaging; it is the factor that produces attitudes, conscious attitudes towards certain phenomena. So when you know, for instance, that there are people who see the difference between red and green, you can take it for a fact that everybody sees that difference. Not at all. There are cases of trichromatism. And so on, you know—the one sees this, the other sees that, and I tried to find out what the principal differences were. That is the book about the types. I saw the introverted and extroverted attitudes, then the functional aspects, then which of the four functions is predominant.

Text pages 103–104

Now mind you, these four functions were not a scheme I had invented and applied to psychology. On the contrary, it took me quite a long time to discover that there is another type—the thinking type, as I thought my type is. Of course that is human. It is not. There are other people who decide the same problems I had to decide in an entirely different way. They look at things in an entirely different way; they have entirely different values. They are, for instance, feeling types. And so, after a while I discovered that there are intuitive types. They gave me much trouble. It took me over a year to become somewhat clear about the existence of the intuitive types. And the last, and the most unexpected, was the sensation type. And only later I saw that those are naturally the four aspects of conscious orientation. You see, you get your orientation, you get your bearings in the chaotic abundance of impressions by the four functions, four aspects; so, if you can tell me any other aspect by which you get your orientation I'm very grateful. I haven't found more. I tried. But those are the four that covered the thing. For instance, the intuitive type, which is very little understood, has a very important function because he is the one going by hunches, he sees around corners, he smells a rat a mile away. He can give you perception and orientation in a situation where your senses, your intellect and your feeling are no good at all. When you are in an absolute fix, an intuition can show you the hole through which you can escape. This is a very important function under primitive conditions, or wherever you are confronted with vital issues. You cannot master by rules or logic or so. So, through the study of all sorts of human types I came to the conclusion that there must be at least as many different worlds—the aspect of the world is not one, it is many, it is at least sixteen, and you can just as well say 360. You can increase the number of principles, but I found the most simple way is the way I told you, the division by four is the simple and natural division of the circle. I didn't know the symbolism of this particular classification. Only when I studied the archetypes I became aware that this is a very important archetypal pattern that plays an enormous role. Also, I found in the study of the types that it gives us a certain lead as to the personal nature of the unconscious, the personal quality of the unconscious in a given case. If you take an extrovert; well, his unconscious has then an introverted quality, because all the extroverted qualities are played in consciousness and the introverted work in the unconscious; therefore it has introverted qualities, etc. With other functions it is the same. That gave me a lead of diagnostic value. It helped me to understand my patients. When I knew their conscious type I got an idea about their unconscious attitude. And since a neurotic is equally influenced by the unconscious, as he is in the conscious, he is influenced by another type, as it were; as if he were another type, and in certain cases it is almost impossible to say whether the individual is to be judged by his conscious quality or by his unconscious quality, because you cannot tell at first sight which is which. So, this has helped me to understand the Freudian aspect and the Adlerian aspect. And then it gave me an important light, also, for the way an individual is going when it is under actual analytical treatment. Because there the point is that you try to integrate unconscious contents into consciousness, or to confront the individual with a definite conscious attitude with its unconscious content that counteracts himself in his neurosis. It is just as though another personality of the opposite type were influencing him, or disturbing him. And so, I got in the course of years quite a great empirical material about the peculiar way in which the conscious and unconscious contents interact.

Text pages 126–132

DR. EVANS: And so, Professor Jung, you began to see in a sense, your typologies led to a sort of theory of psychology of opposites, that the conscious would reveal one side of a type, and the unconscious would be the other side. This would be a very important way, then, of helping you to analyze and understand the individual.

DR. JUNG: Diagnostically, from a practical point of view, it is quite important. The point I wanted to elucidate is—you know, in analyzing a patient you make typical experiences. There is a sort of typical way in which the integration of consciousness takes place. The average way is, that through the analysis of dreams, for instance, you become acquainted with the contents of the unconscious. I have already told you this. To begin with, all personal material: subjective questions, questions of the individual's difficulties in adapting to environmental conditions, etc. Now, it is a regular observation that when you talk to an individual and this individual gives you insight into his inner preoccupations, interests, emotions—in other words, hands over his personal complexes—then you get slowly and willy-nilly into the situation of a sort of authority. You become a point of reference; you know you are in possession of all the important items in a person's development. I remember, for instance, I analyzed a very well-known American politician, and he told me, oh, any amount of the secrets of his trade, and suddenly he jumped up and said, "By God, what have I done! You know, you get a million dollars for what I have told you now!" I said, "Well, I'm not interested. You can sleep in peace, I shall not betray you. I'll forget it within a fortnight." So you see, that shows that the things people hand out are not merely indifferent things. When it comes to something important—emotionally important—then they hand out themselves. They hand out a big emotional value, as if they were handing over a large sum, as if they were trusting you with the administration of their estate, and they are entirely in your hands. Often I hear things that could ruin those people—utterly, permanently ruin—or it would give me, if I should have any blackmailing tendencies, unlimited power to blackmail them. Now you see, that creates an emotional relationship to the analyst, and that is what Freud called the transference, which is a central problem of analytic psychology. It is just so as if these people had handed out their whole existence, and that can have very peculiar effects upon the individual. Either they hate you for it, or they love you for it, but you are not indifferent to them. There is, then, a sort of emotional relation between the patient and the doctor. When you hand out such materials, then these contents are associated with all the important persons in the life of a patient. Now the most important persons are usually father and mother; that comes up from childhood. The first troubles are with the parents, as a rule. So, in handing over your infantile memories about the father or about the mother, you hand over the image of father and mother. Then it is just as if the doctor had taken the place of the father—even of the mother. I have had quite a number of male patients that called me "Mother Jung," because they handed over the mother to me, curiously enough. But you see, that's quite irrespective of the personality of the analyst. It is simply disregarded. It functions as if you were the mother, or it functions as if you were the father—the central authority. Now that is what one calls transference, that is, a typical case of projection. Now Freud doesn't exactly call it projection—he calls it transference. That is an allusion to the old and superstitious idea of handing over disease—transferring the disease upon an animal or handing over the sin upon a scapegoat, and the scapegoat takes it out into the desert and makes it disappear. So they hand over themselves in the hope that I can swallow that stuff and digest it for them. So I am *in loco parentis* and have a high authority. Naturally, I am also persecuted by the correspondent resistances, by all the manifold emotional reactions they have had against their parents. Now that is the structure you have to work through first in analyzing the situation, because the patient in such a condition is not free; he is a slave; he is actually dependent upon the doctor, like a patient with an open abdomen on the operation table. He is in the hands of the surgeon, for better or worse. And so the thing must be finished. And so we have to work through that condition in the hope that we arrive in a situation where the patient is able to see that I am not the father, not the mother—that I am an ordinary human being. Now everybody naturally should assume that such a thing would be possible—that the patient could arrive at such an insight when he is not a complete idiot, that I am just a doctor and not that emotional figure of their fantasies; but that is very often not the case. I had a case, an intelligent young woman; she was a student of philosophy, very good mind, where one could expect easily that she would see that I am not parental authority—but she was utterly unable to get out of this delusion. And in such a case one always has recourse to the dreams. It is just as if one would ask the unconscious, "Now, what do you say to such a condition?" She says through the conscious, "Of course I know you are not my father, but I just feel like that; it is like that. I depend upon you." And I say, "Now we will see what the unconscious says." Now the unconscious produces dreams in which I really assume a very curious role. You know, she was a little infant. She was sitting on my knees. I held her in my arms. I was a very tender father to the little girl, you know, and more and more her dreams became emphatic in that respect, namely, that I was a sort of giant and she is a very little, frail human thing, you know, quite a little girl in the hands of an enormous being. And the last dream of that series was (I cannot tell you all the dreams) was that I—it was out in nature; I stood in a field of wheat, an enormous field of wheat, that was ripe for harvest. I was a giant, and I held her in my arms like a baby, and the wind was blowing over that field of wheat. Now you know, when the wind is blowing over a wheat field it waves, and with these waves I swayed like that putting her, as it were, to sleep. And she felt as being in the arms of a god, of the godhead, and I thought, "Now the harvest is ripe, and I must tell her." And I told her. "You see, what you want and what you project into me because you are not conscious of it is—you have the idea of a deity you don't possess; therefore you see it in me." That clicked. Because, you know, she had a rather intense religious education. Of course, it all vanished later on and something disappeared from her world. The world became merely personal, and that religious conception of the world was nonexistent, apparently. But you see, the idea of a deity is not an intellectual idea. It is an archetypal idea. You find it practically everywhere under this or another name. You know it has the name Mana—it is an all-powerful, extraordinary, effect or quality, even if it is not personal at all. And so she suddenly became aware of an entirely heathenish image that comes fresh from the archetype. She had not the idea of a Christian God, or of an Old Testament Yahweh. It was a heathenish God, you see —a God of nature, of vegetation. He was the wheat himself. He was the spirit of the wheat—the spirit of the wind, and she was in the arms of that numen. Now that is the living experience of an archetype. That made a tremendous impression upon that girl, and instantly it clicked. She saw what she really was missing, that missing value that was in the form of a projection in myself and made myself indispensable to her. And then she saw it is not indispensable, because it is as the dream says she is in the arms of that archetypal idea. That is a numinous experience, you see, and that is the thing that people are looking for—an archetypal experience —that is, then an incorruptible value. They depend upon other conditions; they depend upon their desires, their ambitions. They depend upon other people, because they have no value in themselves. They have nothing in themselves. They are only rational and are not in possession of a treasure that would make them independent. But when that girl can hold that experience, then she doesn't depend anymore. She cannot depend anymore, because that value is in herself. And that is a sort of liberation, and that is, of course, makes her complete, inasmuch as she can realize such a numinous experience, she is able to continue her part—her way, her individuation. The acorn can become an oak, and not a donkey. Nature will take her course. She will become that which she is from the beginning.

Text pages 66–67

Now, having seen such cases—quite a number of such cases —that, of course, has given me a motive to study the archetypes, because I began to see that the structure of what I then called the collective unconscious is really a sort of agglomeration of such typical images, of which each has a numinous quality. The archetypes are, at the same time, dynamic; they are instinctual images that are not intellectually invented. They are always there and they produce certain processes in the unconscious, one could best compare with myths. That's the origin of mythology. Mythology is a pronouncing of a series of images that formulate the life of archetypes. And so the statements of every religion, of many poets, etc., are statements about the inner mythological process, which is a necessity, because man is not complete if he is not conscious of that aspect of things.

Text pages 152–155

And so, you see, man is not complete when he lives in a world of statistical truth. He must live in a world of his biological truth, that is his biological truth—that is not merely statistics. It is the expression of what he really is, and what he feels himself. So you see, someone without the mythology is merely an effect of statistics, as it were. He is an average phenomenon. And while the truth is, the carriers of life are individuals, not average numbers, yet our natural science makes everything to an average—reduces everything to an average. And of course, all the individual qualities are wiped out. That, of course, is most unbecoming. It is unhygienic. It deprives people of their specific values. Where they are individuals, it deprives them of the most important experiences of their life, where they experience their own value—the creative background of their personality. You see, the trouble is that nobody understands these things, apparently. It is quite strange.

DR. EVANS: You think that the humanities are important for the individual who wants to study the individual?

DR. JUNG: One doesn't see what an education without humanities is doing to man. He loses the connection with his family, as it were—the whole stem, the tribe—the connection with the past, that he lives in, that in which man always has lived. Man has always lived in the myth, and we think we are able to be born today and to live in no myth—without history. That is a disease—absolutely abnormal—because man is not born every day. He is once born in a specific historical setting with specific historical qualities, and therefore he is only complete when he has a relation to these things. It is just as if you were born without eyes and ears, when you are growing up with no connection with the past. From the standpoint of natural science, "You need no connection with the past; you can wipe it out," and that is a mutilation of the human being. Now I saw from a practical experience that this kind of proceeding has a most extraordinary therapeutical effect. I can tell you such a case. There was a Jewish girl. Her father was a banker, and she had received an entirely worldly education; she had no idea of any tradition, also, and then I went further into her history and found out that her grandfather had been a Saddik in Galicia, and when I knew that, I knew the whole story. That girl suffered from phobia—a terrific phobia, and had been under psychoanalytic treatment already with no effect, and she was really badly plagued by that phobia, in excited states and so on, and then I saw that girl has lost the connection with her past—has lost, for instance, the fact that her grandfather was a Saddik—that he lived in the myth. And her father has fallen out of it, too. So, I simply told her, "You will stand up to your fear, you know what you have lost," and they didn't, of course not. I said, "Your fear is the fear of Yahweh." You know, the effect was that within a week she was cured from so many years of bad anxiety states because, you see, that went like lightning through her. I only could say it because I knew that she is absolutely lost. She thought she was in the middle of things, but she was lost, gone. It made no sense—what is our existence when we are just average numbers? The more you make people into average numbers the more you destroy our society. The "ideal state" and the "slave state." Go to Russia. There it is wonderful; there you can be a number, but one pays very dearly; our whole life goes to blazes. And so, you see, I have plenty of cases of a similar kind. And that has, naturally, led me on to a profound study of the archetype; I got more and more respectful of archetypes. And now, by Jove, that thing should be taken into account.

Text pages 67–70

That is an enormous factor, very important for our further development and for our well-being. It was, of course, difficult to know where to begin, because it is such an enormously extended field. And the next question I asked myself was, "Now, where in the world has anybody been busy with that problem?" And I found nobody except a peculiar spiritual movement that went together with the beginning of Christianity, namely, the Gnostics. And that was the first thing, actually, that I saw. They were concerned with the problem of archetypes, and made a peculiar philosophy of it—and everybody makes a peculiar philosophy of it when he comes across it naively, and doesn't know that those are structural elements of the unconscious psyche. Now they have lived in the first, second, and third century. And what was in between? Nothing. And now, today, we suddenly fall into that hole and are confronted by the problems of the collective unconscious which were then the same, 2,000 years ago, and we are not prepared to admit that problem. I was always looking for something in between, you know—something that links that remote past with the present moment. And I found to my amazement it is ˙alchemy, that is understood to be a history of chemistry. It is, one could almost say, anything but that. It is a peculiar spiritual movement or a philosophical movement. They called themselves philosophers, like Gnosticism. And then I read the whole accessible literature, Latin and Greek. I studied it because it was enormously interesting. It is the mental work of 1,700 years in which there is stored up all they could make out about the nature of the archetypes, in a peculiar way, that's true, it is not simple. And the most of the texts are no more published since the Middle Ages; the last editions, in the middle or end of the sixteenth century—all in Latin; some texts are in Greek—not a few very important ones. That has given me no end of work, but the result was most satisfactory, because it showed me the development of our unconscious relations to the collective unconscious and the variations our consciousness has undergone; why the being is unconscious is concerned with these mythological images. For instance, such phenomena as in Hitler; you know, that is a psychical phenomenon, and we've got to understand these things. It is just as though—as if a terrific epidemic of typhoid fever were breaking out, and you say, "That is typhoid fever— isn't that a marvelous disease?" It can take on enormous dimensions and nobody knows anything about it. Nobody takes care of the water supply, nobody thinks of examining the meat or something like that, but simply states that this is a phenomenon; yes, but one doesn't understand it. And to me, of course, it was an enormous problem because it is a factor that has determined the fate of millions of European people, and of Americans. Nobody can deny that he has been influenced by the war. That was all Hitler's doing—and that's all psychology, our foolish psychology. But you only come to an understanding of these things when you understand the background from which they spring. And so, of course, I cannot tell you in detail about alchemy. It is the basis of our modern way of conceiving things and therefore it is as if it were right under the threshold of consciousness. This is a wonderful picture of how the development of archetypes—the movement of archetypes looks—when you look upon them as if from above. Maybe from today you look back into the past and you see how the present moment has evolved out of the past. It is just as if the alchemical philosophy—it sounds very curious; we should give it an entirely different name. It has a different name. It is called Hermetic philosophy. Of course, that conveys just as little as the term alchemy. It is the parallel development as Gnosticism was to the conscious development say of Christianity—of our Christian philosophy—of the whole psychology of the Middle Ages. And so you see, in our days we have such and such a view of the world—such a philosophy—but in the unconscious we have a different one. And that we can see through the example of the alchemical philosophy that behaves towards us, the medieval consciousness exactly like the unconscious behaves to ourselves. And we can construct or even predict the unconscious of our days when we know what it has been yesterday. Now that is, in a few words, the development of my ideas, you see, without going into detail.

Text page 159

DR. EVANS: But you've gone into great detail elsewhere in much of your writing, of course.

DR. JUNG: Well, people have to read the books, by golly, in spite of the fact that they are thick. I'm sorry.

DR. EVANS: We are hoping this will stimulate many of them to read the books.

DR. JUNG: Well, you can stop right here.

DR. EVANS: Well done! Well done, Dr. Jung.

REFERENCES

ADLER, A. *The Practice and Theory of Individual Psychology.*
New York: Harcourt, 1927.

EVANS, R. I., Roney, H. B., McAdams, W. J. "An Evaluation of
the Effectiveness of Instruction and Audience Reaction to
Programming on an Educational Television Station."
Journal of Applied Psychology 39 (1955) 277–279.

————. "Contributions to the History of Psychology: Filmed
Dialogues with Notable Contributors to Psychology."
Psychological Reports, 1969c, 25, 159–164.

————. "A Conversation with Carl Jung," *Psychology Today,*
1967, 1, 3, 35–39.

FREUD, S. *Group Psychology and the Analysis of the Ego.*
Translation, London: International-Psychoanalytical Press,
1922.

————. *Beyond the Pleasure Principle.* Translation, London:
International Psychoanalytic Press, 1922.

————. *The Interpretation of Dreams.* Translation by A. A. Brill. New York: Modern Library, 1950.

HALL, C. S. & LINDZEY, G. *Theories of Personality.* 2d. ed. New York: John Wiley & Son, Inc., 1970.

HITE, H. "A Study of Teacher Education Methods for Audiovisual Competency in Washington 1937–1947." Unpublished doctoral dissertation, State College of Washington, 1951.

HEIDEGGER, M. *An Introduction of Metaphysics.* Translation by R. Manheim. New Haven, Conn.: Yale University Press, 1959.

HORNEY, K. *The Neurotic Personality of Our Time.* New York: W. W. Norton & Company, 1937.

HUSSERL, E. *Ideas: General Introduction of Pure Phenomenology.* Translation by W. R. Boyce Gibson. New York: The Macmillan Company, 1952.

JUNG, C. G. *Symbols of Transformation.* (A revised edition of *The Psychology of the Unconscious.*) Vol. 5. *Collected Works.* New York: Pantheon Press, 1956.

————. *Memories, Dreams, and Reflections.* Recorded and edited by Aniela Jaffé. Translation by R. C. Winston. New York: Pantheon Books, 1963.

KUMATA, H. "An Inventory of Instructional Television Research." Ann Arbor: Educational Television and Radio Center, 1956.

LEWIN, K. *Principles of Topological Psychology.* New York: McGraw, 1936.

MASLOW, A. H. *Motivation and Personality.* New York: Harper, 1954.

MAY, R. *The Meaning of Anxiety.* New York: Ronald, 1950.

MCCURDY, H. G. *The Personal World: an Introduction to the Study of Personality.* New York: Harcourt, Brace & World, 1961.

MILLER, N. E. "Graphic Communication and the Crisis in Education." Washington Educational Association, 1957.

OSGOOD, C. E. "The Measurement of Meaning." *Psychological Bulletin,* 1952, 49, 197–237.

PAULI, W. "Influence of Archetypal Ideas on the Scientific Theories of Kepler." In Jung, C. G., and Pauli, W., *The Interpretation and Nature of the Psyche.* New York and London: Pantheon Books, 1955.

RANK, O. *Will Therapy.* Translation by Julia Taft. New York: Knopf, 1945.

ROCK, R. T., JR., DUVA, J. S. & MURRAY, J. E. "Training by Television: the Comparative Effectiveness of Instruction by Television, Television Recordings, and Conventional Classroom Procedures." Port Washington, L.I., New York: Special Devices Center, 1957. (SDC Rep. 476–02–2 [Navexos pp. 850–852].)

INDEX

ABOUT THE AUTHOR

RICHARD I. EVANS received his Ph.D. from Michigan State University and is currently professor of psychology at the University of Houston. He is the Director of the Social Psychology/Behavioral Medicine Research and Graduate Training Group.

A National Science Foundation grant has enabled him to film discussions and complete books with some of the world's foremost behavioral scientists including the distinguished participants in the dialogues in this Praeger Series.

He is a pioneer in public television and in the social psychology of communication, and has published over a hundred articles in the area of social psychology. In addition to the volumes in this Dialogue Series, his books include *Social Psychology in Life* (with Richard Rozelle), *Resistance in Innovation in Higher Education, The Making of Psychology,* and *The Making of Social Psychology.*

His recent honors include the American Psychological Foundation Media Awards for the book, *Gordon Allport: The Man and His Ideas* and the film, "A Psychology of Creativity." He and his colleagues received American Psychological Association Division 13 Research Excellence Awards in 1970, 1973, and 1977 for their work in social psychology in behavioral medicine. He received the Phi Kappa Phi National Distinguished Scholar Award for the 1974-77 Triennium, and the 1980 Ester Farfel Award, the University of Houston's highest award for excellence in teaching, research and service.